Christina Stawiarski

Optimizing Dual-Doppler Lidar Measurements of Surface Layer Coherent Structures with Large-Eddy Simulations

Wissenschaftliche Berichte des Instituts für Meteorologie und
Klimaforschung des Karlsruher Instituts für Technologie (KIT)
Band 64

Herausgeber: Prof. Dr. Ch. Kottmeier

Institut für Meteorologie und Klimaforschung
am Karlsruher Institut für Technologie (KIT)
Kaiserstr. 12, 76128 Karlsruhe

Eine Übersicht über alle bisher in dieser Schriftenreihe erschienenen Bände
finden Sie am Ende des Buches.

Optimizing Dual-Doppler Lidar Measurements of Surface Layer Coherent Structures with Large-Eddy Simulations

by
Christina Stawiarski

Dissertation, Karlsruher Institut für Technologie (KIT)
Fakultät für Physik, 2014
Referent: Prof. Dr. Christoph Kottmeier
Korreferent: Prof. Dr. Siegfried Raasch

Impressum

 Scientific Publishing

Karlsruher Institut für Technologie (KIT)
KIT Scientific Publishing
Straße am Forum 2
D-76131 Karlsruhe

KIT Scientific Publishing is a registered trademark of Karlsruhe
Institute of Technology. Reprint using the book cover is not allowed.

www.ksp.kit.edu

Print on Demand 2014

ISSN 0179-5619
ISBN 978-3-7315-0197-8
DOI: 10.5445/KSP/1000039726

Optimizing Dual-Doppler Lidar Measurements of Surface Layer Coherent Structures with Large-Eddy Simulations

Zur Erlangung des akademischen Grades eines

DOKTORS DER NATURWISSENSCHAFTEN

der Fakultät für Physik

des Karlsruher Instituts für Technologie (KIT)

genehmigte

DISSERTATION

von

Christina Stawiarski

aus Hagen

Tag der mündlichen Prüfung: 7. Februar 2014

Referent: Prof. Dr. Christoph Kottmeier

Korreferent: Prof. Dr. Siegfried Raasch

Zusammenfassung

Kohärente Strukturen sind zusammenhängende Bereiche in einem Fluid, in denen Variablen wie die Komponenten des Windfeldes oder die Temperatur eine hohe Korrelation mit sich oder anderen Variablen aufweisen. Diese Gebiete, welche deutlich größer sind als die kleinsten Skalen der Turbulenz, sind als regelmäßige Muster im Windfeld der atmosphärischen Grenzschicht erkennbar. Die Strukturen tragen im hohem Maße zu turbulenten Flüssen bei und beeinflussen so Transport und Durchmischung in der Grenzschicht. Nahe der Erdoberfläche können kohärente Strukturen mit Längenskalen von 100 m bis zu wenigen Kilometern beobachtet werden.

Doppler-Lidare sind aktive Fernerkundungsinstrumente, die besonders für die Untersuchung atmosphärischer Windfelder in der Grenzschicht geeignet sind. Der Einsatz zweier synchron gesteuerter Lidare ermöglicht die Vermessung des horizontalen Windfeldes in Bodennähe auf Gebieten von der Größe mehrerer Quadratkilometer mit hoher räumlicher und zeitlicher Auflösung. Obwohl Doppler-Lidare damit für die Untersuchung kohärenter Strukturen optimal geeignet scheinen, wird ihre Detektion durch die im Messverfahren inhärenten räumlichen und zeitlichen Mittelungsprozesse erschwert. Es ergibt sich die Frage: Wie gut können kohärente Strukturen mit dem Dual-Doppler Lidar Verfahren detektiert und vermessen werden?

Um diese Frage zu beantworten werden hochaufgelöste Grobstruktursimulationen der Grenzschicht als Grundlage für virtuelle Lidar-Messungen verwendet: mit Hilfe mathematischer Modelle für Lidar-Messungen wird berechnet, was ein Doppler-Lidar in durch die Grobstruktursimulation vorgegebenen Windfeldern messen würde. Damit ist es möglich, die Effekte des Lidar-Messverfahrens auf die detektierte Struktur des Windfeldes direkt sichtbar zu machen und zu analysieren. Auf die Grobstruktursimulationsdaten und die virtuellen Doppler-Lidar-Daten werden bekannte Strukturdetektionstechniken angewandt: die Bestimmung der räumlichen integralen Längenskala, eine Wavelet-Analyse sowie die Gruppierung zusammenhängender Bereiche geringer Windgeschwindigkeit. Mittels theoretischer Betrachtungen und einer Fehleranalyse werden außerdem Techniken zur Optimierung von Dual-Doppler Lidar Verfahren und zur Bewertung und Korrektur der gemessenen Struktur-Längenskalen entwickelt.

Qualitativ ergibt sich, dass große Strukturen ($> 10\Delta xy$) zuverlässig detektiert werden können, die Größe kleinerer Strukturen (< 1 bis $5\Delta xy$) jedoch überschätzt wird, wobei die Strukturgröße im Verhältis zur Lidar-Auflösung Δxy zu betrachten ist.

Die quantitative Analyse zeigte, dass die integrale Längenskala nach Korrektur ein geeignetes Maß zur Bestimmung der räumlichen Korrelationslänge aus Lidar-Daten darstellt. Die Wavelet-Analyse eignet sich zur Untersuchung einzelner Strukturen nur dann, wenn diese eine Mindestgröße von 5 bis 10 Δxy überschreiten welche durch die Lidar-Auflösung festgelegt wird. Die Gruppierungsmethode ist für die vorliegenden Lidar-Daten ungeeignet.

Die hier erarbeiteten theoretischen Ergebnisse werden in der Durchführung und Auswertung des HOPE-Experiments im Frühjahr 2013 mit den KIT Lidar-Systemen angewandt.

Abstract

Coherent structures are connected regions in a fluid in which variables like wind speed or temperature exhibit a high correlation with themselves or other variables. These regions, which are much larger than the smallest scales of turbulence, appear as regular patterns in the wind field of the atmospheric boundary layer. The structures account for a large proportion of turbulent fluxes and thereby influence atmospheric boundary layer mixing and transport. Close to the earth's surface the structures on length scales of 100 m to few kilometers can be observed.

Doppler lidars are active remote-sensing instruments for atmospheric wind measurements in the boundary layer. The deployment of two synchronously scanning ground-based lidars facilitates the measurements of the horizontal wind field close to the surface on an area of several square kilometers with unprecedented time and spatial resolution.

Although Doppler lidars appear ideally suited to investigate these structures, the inherent averaging processes involved in lidar measurements complicate the structure detection. The question arises: How well do Doppler-lidars perform in the detection and measurement of coherent structures?

To answer this question high-resolution large-eddy simulations of the boundary layer are used as a basis for virtual dual-lidar measurements. This way it becomes possible to directly visualize and analyze the effects of the lidar measurement technique on the detected wind field structure. Three common structure detection techniques are applied to both the large-eddy simulation data and the virtual lidar scan data: the computation of spatial integral length scales, a wavelet-analysis, and the clustering of low wind-speed regions. Based on theoretical investigations and an error discussion techniques are developed to optimize dual-lidar scans and to assess and correct the measured coherent structure length scales.

The qualitative evaluation reveals that large structures ($> 10\,\Delta xy$) can be detected reliably, whereas the size of smaller structures (< 1 to $5\,\Delta xy$) is overestimated. Structures are seen as large or small relative to the lidar resolution Δxy. The quantitative analysis shows that the integral length scales derived from lidar-data are after a correction suitable to determine spatial correlations lengths. The wavelet-analysis is best suited for the investigation of single structures, provided that these exceed a length scale of 5 to 10 Δxy determined by the lidar resolution. The clustering algorithm is unsuitable for the application on the present dual-lidar data. The theoretical results developed in this study were applied during the KIT dual-lidar system deployment and the subsequent data analysis in the HOPE-experiment in spring 2013.

Contents

1. Introduction

Atmospheric turbulence describes the random and chaotic motion of air in elements ('eddies') of different scales over several orders of magnitude, which form, interact, and decay. Hinze (1959) defines: "Turbulent fluid motion is an irregular condition of flow in which the various quantities show a random variation with time and space coordinates, so that statistically distinct average values can be discerned." It is created through wind shear and buoyancy, which become particularly relevant near the earth's surface. Turbulent mixing is the most important transport process in the atmospheric boundary layer, i.e. the part of the atmosphere which is influenced by the earth's surface, with a height of up to 3 km. Although it is generally agreed that the Navier-Stokes-Equations describe turbulent motions, the mathematical problems they pose, which are caused by their non-linearity, remain unsolved. In fact, a proof of the existence and smoothness of a solution has been posed as one of the seven 'Millennium Problems' (Carlson et al., 2006).

The stochastic nature of turbulence in fluids has been studied for more then 100 years under laboratory and free atmospheric and oceanic conditions.
Recurrent coherent structures are rather regular patterns in fully developed turbulent flow fields, such as in the atmospheric boundary

layer.[1] In the surface layer, which is the boundary layer region close to the ground, these structures are responsible for a large, possibly the dominant, part of the fluxes of momentum and heat. In the shear-driven surface layer, they appear as elongated streaks of alternatingly low- and high-speed fluid which coincide with up- and downdrafts, respectively, and which become the starting points for horizontal convective rolls in the sheared convective boundary layer. Under conditions of free convection, the updraft regions arrange in regular, hexagonal patterns. The surface layer structures occur on length scales from less than 100 meters to a few kilometers. Although they approach the resolution of mesoscale forecast models like COSMO-DE, these structures are not considered in the sub-filter-scale parameterization. During several decades of research, these structures have been investigated with atmospheric measurements, wind tunnel measurements, and numerical models. However, there is still no conclusive knowledge about their scales, intensities, and contributions to fluxes and turbulent kinetic energy, which have been found to depend on shear, stability, boundary layer height, surface roughness, and heterogeneity.

A main difficulty is the challenge of capturing the large-scale structures with meteorological instruments: For years, point measurements from towers and aircrafts were the sole source of data on coherent structures in the atmospheric boundary layer. Recently, remote sensing instruments have become available for coherent structure research: Doppler lidars are, depending on their specifications, able to measure the wind component in beam direction with a resolution of about 50 to 100 meters over a range of ten to fifteen kilometers. Compared to Doppler radars, which only detect liquid water content, Doppler lidar pulses have

---

[1]Various more detailed definitions of the term 'coherent structures' can be found in Chap. 2.2.

a wavelength in the infrared which is back-scattered by atmospheric aerosols moving with the wind. Lidars are therefore well suited for wind speed measurement in the boundary layer.

Two Doppler lidars scanning the surface layer synchronously are able to retrieve the complete horizontal wind field on an area of several square kilometers. This approach has first been applied by Rothermel et al. (1985). Using this method, the structures in the surface layer can been detected. However, the scanning time and the spatial averaging involved in lidar measurements are important error sources whose influence on structure detection has not yet been investigated. This advanced measurement technique therefore requires an assessment of its accuracy. Comparative measurements are complicated by the fact that no other instruments provide data with comparable range and resolution.

Increasing computational power has lead to a considerable improvement in boundary layer modeling. Large-eddy simulations (LES) are able to resolve turbulence down to scales of a few meters and have been widely used to investigate boundary layer structures. However, LES analyses must be accompanied by measurements since the smallest scales of turbulence and the region close to the surface, which are particularly important for the structure generation, have to be parameterized.

This study has the objective to determine the quality of coherent structure detection techniques in dual-Doppler surface layer scan data. To this effect, surface layer wind fields from LES are employed which exhibit coherent structures. Virtual dual-Doppler lidar scans and retrievals are performed based on the LES fields with a lidar simulation tool developed as a part of this work. The lidar simulator yields the radial wind velocities that a real lidar would 'see' in the LES atmosphere. By comparing the structures determined with different techniques from the 'real' LES, including virtual tower measurements, and the 'measured' virtual lidar data, the quality of the techniques are assessed. Where possible, correction techniques are developed based on the mathematical model of lidar measurements.

With these results, real dual-Doppler lidar measurements can be interpreted and the present coherent structures can be characterized including error estimates.

In the HOPE experiment (HD(CP)^2 Observational Prototype Experiment), which was part of the HD(CP)^2 campaign in Jülich in spring 2013 (High Definition Clouds and Precipitation for Advancing Climate Prediction)[2], the KIT dual-Doppler lidar system performed low-elevation surface-layer scans, using a synchronized control system and an optimized scanning pattern developed as a part of this study. The analyses of this work will be used to evaluate and correct the structure scale of the experimental results. An outlook on is given in the conclusion.

This work is organized as follows:

In Chap. 2, the concept of coherent structures in turbulent fluids is introduced, and the current state of research about atmospheric structures from both tower and remote sensing data as well as LES is summarized. Chap. 3 gives an overview of single and dual-Doppler lidar measurements in the boundary layer, as well as the lidar simulation tool used for virtual measurements based on LES data. Data from real and virtual dual-lidar measurements are reassembled using a retrieval algorithm also introduced here.

---

[2]Until January 2014 no articles on the HD(CP)^2 experiment were published, but an overview of the experiment could be found at the website http://hdcp2.zmaw.de .

In Chap. 4 the various error sources and their influence on dual-Doppler measurements are discussed, which leads to an optimization algorithm for scanning patterns. This optimization is applied in Chap. 5, where virtual dual-Doppler measurements and retrievals are performed on four LES data sets with varying shear and convective forcings.

In Chaps. 7 and 6, the high-resolution LES data and the virtual dual-lidar retrieval data are both evaluated with the same four coherent structure detection techniques. While in Chap. 6 length scales in the horizontal wind field are investigated using correlation lengths, a wavelet algorithm and a clustering approach, in Chap. 7 structures in the vertical wind are detected using a Lagrangian Coherent Structure algorithm. The various methods are compared with respect to their applicability with dual-Doppler data and where possible correction methods are developed and tested.

Finally, Chap. 8 summarizes the results and gives recommendations for dual-Doppler lidar scan and evaluation techniques for coherent structure detection. Additionally, first results from the HOPE experiment are presented.

2. Coherent Structures in the Atmospheric Boundary Layer

This chapter gives an overview of coherent structure research, reaching from their discovery and visualization in low Reynolds number flows of wind tunnels to their investigation in atmospheric boundary layers.
The current knowledge about atmospheric coherent structures and their relation to boundary layer scaling parameters was shaped by numerical simulations using LES, and measurements with meteorological towers and remote sensing instruments.

2.1. Scaling Parameters in the Atmospheric Boundary Layer

The atmospheric boundary layer is a turbulent fluid in which turbulent kinetic energy (TKE) is generated from buoyant and shear forcing. This is expressed by the first two terms on the right hand side of the TKE-equation (Stull, 1988):

$$\frac{\partial \bar{e}}{\partial t} + \bar{u}_j \frac{\partial \bar{e}}{\partial x_j} = \delta_{i3} \frac{g}{\theta_v} \left(\overline{u_i' \theta_v'} \right) - \overline{u_i' u_j'} \frac{\partial \bar{u}_i}{\partial x_j} - \frac{\partial \overline{u_j' e}}{\partial x_j} - \frac{1}{\rho} \frac{\partial \overline{u_i' p'}}{\partial x_i} - \varepsilon \,, \qquad [2.1]$$

where $\bar{e} = \frac{1}{2}\overline{u_i'^2}$ is the turbulent kinetic energy, u_i the ith wind field component $\{u_1, u_2, u_3\} = \{u, v, w\}$, g the gravitational constant, θ_v the virtual potential temperature, ρ the density of air, p the pressure and ε the dissipation. Summation over repeated indices is assumed, δ_{ij} is the

Kronecker symbol. Here, $\bar{x}$ denotes the average of a variable x, and x' its deviation from the average: $x' = x - \bar{x}$.

To estimate the relative influence of shear and buoyancy, the friction velocity u_* and the convective velocity scale w_* are introduced as scaling parameters (Deardorff, 1972; Stull, 1988):

$$u_* = \left(\overline{u'w'}_0^2 + \overline{v'w'}_0^2\right)^{1/4} ,$$

[2.2a]

$$w_* = \frac{g}{\theta_v}\left(\overline{w'\theta'_v}_0 \cdot z_i\right)^{1/3} .$$

[2.2b]

Here, z_i is the boundary layer height, and the index 0 denotes the values at the surface.

The scaling parameters are used as the relevant scales for the dimensionless groups in similarity theory (Buckingham, 1914).
In the lower part of the boundary layer, about the lowest 10%, the fluxes are approximately constant. In this part, called the surface layer, Monin-Obukhov similarity is assumed when considerable shear is available. The relevant scales for length and wind speed given are by the Obukhov-length L_*, u_*, and the roughness length z_0. It is assumed that the boundary layer height does not affect the surface layer.
On the other hand, in the mixed layer with calm or light winds, shear becomes irrelevant and mixed-layer similarity is assumed, where lengths and wind speeds scale with z_i and w_* (Stull, 1988).
The Obukhov-length L_* is defined as

$$L_* = -\frac{\left(\overline{u'w'}_0^2 + \overline{v'w'}_0^2\right)^{3/4}}{\frac{g}{\theta_v}\overline{w'\theta'_v}_0 \cdot \kappa} ,$$

[2.3]

where $\kappa \approx 0.4$ is the von Kármán constant. One physical interpretation of L_* is that it is a measure for the height above ground at which buoyancy first dominates of shear (Stull, 1988).

As a measure for stability, the ratio

$$-\frac{z_i}{L_*} = \kappa \frac{w_*^3}{u_*^3} \qquad [2.4]$$

is commonly used. When $-\frac{z_i}{L_*} > 0$, buoyant production exceeds buoyant consumption at the surface and the boundary layer is convective and unstable. When $-\frac{z_i}{L_*} > 1$, buoyant production exceeds shear production. For larger $-z_i/L_*$ the flow becomes increasingly unstable. On the other hand, negative $-z_i/L_*$ means an excess of buoyant consumption, and the stratification becomes stable.

In comparisons with laboratory experiments, the Reynolds number becomes another important scaling parameter:

$$Re = \frac{\overline{U}\,\overline{L}}{\nu} . \qquad [2.5]$$

Here, $\overline{U}$ and $\overline{L}$ are the velocity and length scale of the flow, respectively, and ν is the kinematic viscosity.

The Reynolds number is a measure for the range of scales between the largest and the smallest turbulent elements ('eddies') (Fröhlich, 2006, Chap. 2):

$$\frac{L}{\eta} = Re^{3/4} , \qquad [2.6]$$

where L is the largest scale of turbulence, and $\eta = \left(\nu^3/\varepsilon\right)^{1/4}$ is the Kolmogorov microscale which denotes the size of the smallest eddies. Turbulence spectra show that the large eddies carry the largest proportion of energy. This energy is transported down across the inertial range to the smallest scales of the spectrum, where it is finally dissipated into heat. While in laboratory settings low Reynolds numbers

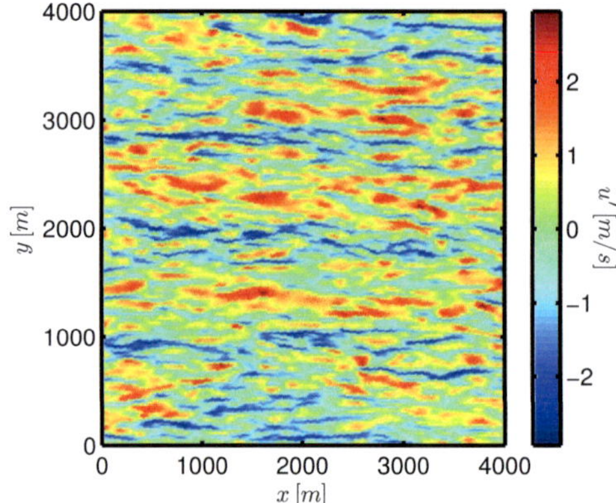

Figure 2.1: The streamwise turbulent wind field component u' in the horizontal plane (x is the mean-wind direction) at $z = 10$ m from an LES with the PALM model at $-z_i/L_* = 1.9$.

can be achieved ($Re < 1000$), the atmospheric boundary layer exhibits $Re > 10^7$, meaning that five orders of magnitude lie between the smallest and largest turbulent scales. This is a challenge for turbulence research in the atmospheric boundary layer, which can only partly be overcome by the application of similarity theories.

2.2. Organized Motions in Wall-Bounded Turbulent Flows

Several decades ago it was discovered in laboratory experiments with low-Reynolds number flows that turbulent shear flows close to the surface exhibit ordered structures in the wind field and other variables (Grant, 1958; Kline et al., 1967): The streamwise turbulent wind component, u', showed spatially coherent regions with $u' < 0$, which were elongated in the direction of the mean wind and which alternated in spanwise direction with regions of enhanced wind velocity, $u' > 0$. These regions were denoted as 'streaks' or 'streaky structures'. They appear also in high-Reynolds-number fluids like the atmospheric surface layer, as shown in the large-eddy simulation of Fig. 2.1. The streaks feature

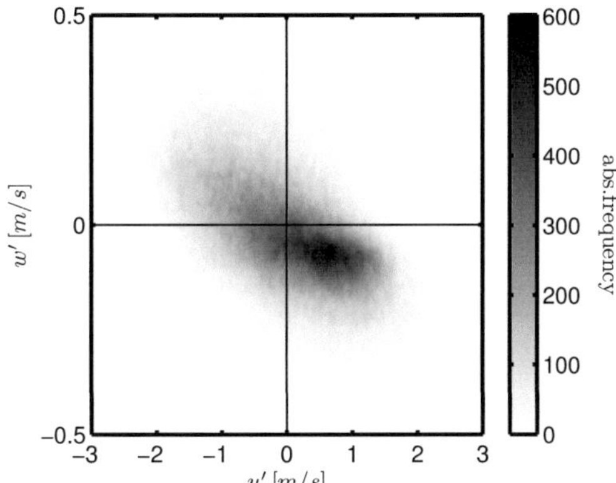

Figure 2.2: Correlation of u' and w' at $z = 10$ m for the situation in Fig. 2.1.

an anticorrelation between u' and w' (cf. Fig. 2.2), therefore the regions with $u' > 0$ and $w' < 0$ are commonly called sweeps, whereas regions with $u' < 0$ and $w' > 0$ are called ejections.

Streaks are only one manifestation of what is generally called 'coherent structures' or 'coherent motions'. These terms are used to describe repetitive patterns in the boundary layer variables, although the scientific community has not yet agreed on what exactly constitutes a coherent structure and the vortices often associated with the structures (Robinson, 1991; Mathieu and Scott, 2000, Chap. 5.5). Common definitions are either too vague to allow comparable quantitative analyses, inextricably linked with a certain detection method, or require highly resolved information on one or more variable of the fluid over a certain volume. Below, some exemplary definitions are summarized. Furthermore, Chakraborty et al. (2005) compare various local vortex identification schemes which are used for the definition of structures.

Hussain (1983) "A coherent structure is a connected, large-scale turbulent fluid mass with a phase-correlated vorticity over its spatial extent. [...] The largest spatial extent over which there is coherent vorticity denotes the extent of the coherent structure."

Robinson (1991) "[...] a coherent motion is defined as a three-dimensional region of the flow over which at least one fundamental flow variable (velocity component, density, temperature, etc.) exhibits significant correlation with itself or with another variable over a range of space and/or time that is significantly larger than the smallest local scales of the flow."

Jeong and Hussain (1995) "Turbulent shear flows have been found to be dominated by spatially coherent, temporally evolving vortical motions, popularly called *coherent structures*. [...]
$\mathbf{S}$ and $\mathbf{\Omega}$ are the symmetric and antisymmetric components of $\nabla \mathbf{u}$; i.e. $S_{i,j} = \frac{1}{2}\left(u_{i,j} + u_{j,i}\right)$ and $\Omega_{i,j} = \frac{1}{2}\left(u_{i,j} - u_{j,i}\right)$. [...] Thus, we [...] define a vortex core as a connected region with two negative eigenvalues of $\mathbf{S}^2 + \mathbf{\Omega}^2$."

Lin et al. (1996) "[...] 'coherent structures' [...] are recurrent, spatially local flow patterns which are long-lived in a Lagrangian reference frame (i.e., moving with the local fluid velocity) and which have deterministic, chaotic, intermittently dissipative dynamics."

Adrian (2007) "One of the principal schools of thought in the study of turbulence seeks to break the complex, multiscaled, random fields of turbulent motion down into more elementary organized motions that are variously called eddies or coherent structures. These motions can be thought of as individual entities if they persist for long times, i.e., if they possess temporal coherence. By virtue of fluid continuity, all motions possess some degree of spatial coherence, so coherence in space is not sufficient to define an organized motion. Only motions that live long enough to catch our eye in a flow visualization movie and/or contribute significantly to time- averaged statistics of the flow merit the study and attention we apply to organized structures."

Zhang et al. (2011) "In ABL, coherent structures are usually defined as low-frequency, large-scale phase-related organized motions that interact with well-known high-frequency, small-scale turbulence."

Zeeman et al. (2013) "It is generally accepted that vertical transport near the surface exhibits forms of organized motion. These are termed 'coherent structures' and contribute to momentum transfer and transport from within canopies through intermittent, shear-induced gusts or 'sweeps' that in turn cause upward bursts or 'ejections' from the canopy."

Coherent motions manifest as streaks, quasi-streamwise and hairpin-shaped vortices in the surface layer which can move and grow through the mixed layer, as large-scale horizontal roll vortices or hexagonal spoke-patterns in the vertical wind. The latter two examples appear to require convective conditions for their development. Robinson (1991) and Adrian (2007) give reviews on coherent structures in low-Reynolds number wall-bounded flows, streaks and rolls in the atmospheric boundary layer are reviewed in Young et al. (2002).

The ambiguity about coherent structure definitions exacerbates the comparability of quantitative results. It also underlines the fact that the different research fields and groups have different objectives in investigating coherent structures, which reach from the characterization of single vortical elements in shear flows (e.g. Adrian et al., 2000) to the statistical contributions of coherent motions to large-scale turbulent fluxes in the atmosphere (e.g. Kim and Park, 2003).

2.3. Structure Characteristics in Low-Reynolds-Number Flows

Laboratory experiments on fluid dynamics usually study low-Reynolds-number flow in a flat-plate boundary layer without surface heat-flux. Early flow visualizations in laboratory settings reported low-speed streaks in the near-wall region with an approximate spanwise spacing of 100 viscous wall units $\delta_v = v/u_*$ (Kline et al., 1967), and turbulent bulges in the outer layer with length scales of 2-3 z_i (Falco, 1977).
Using conditional sampling techniques, Willmarth and Lu (1972) showed that in the near-wall low-speed regions 'bursting' processes occur, which are intermittent and strong events in which the fluid is ejected away from the wall. These ejections are the dominant source of Reynolds-stress and TKE production in the wall region ($z/\delta_v \leq 100$), followed by the contribution of the reversed motions of sweeps.

Increasing evidence suggested that the near-wall structures are associated with vortical motions, variably called arches, quasi-streamwise vortices, horseshoe-vortices, hairpin-vortices and similar (cf. Robinson, 1991).
When the evolution of computational power allowed large-eddy simulations (LES) and Direct Numerical Simulations (DNS) of the flow, the existence of hairpin-shaped vortices was confirmed in these models

(Moin and Kim, 1985; Kim and Moin, 1986).

Adrian and Liu (2002) used DNS to extract the shape of a 'conditional eddy', which is an ensemble-average of the flow-field around strong ejections. This approach to define a coherent structure has the advantage of being oriented towards determination of large contributions to the Reynolds-stress. The shape of the structure closely resembles a hairpin-vortex with a lifted arch-shaped 'head' and trailing 'legs' consisting of two counter-rotating streamwise vortices. Strong ejections occur inside the hairpin, whereas downstream the vortex induces a less intense sweep motion.

Based on research results up to that point, Adrian, Meinhart, and Tomkins (2000) developed a conceptual model to explain coherent structures in the flat-plate boundary layer:

When a spanwise vortex filament close to the wall is lifted by a random disturbance, the strong shear will lead to a stretching in the mean-flow direction, which then leads to the evolution of the hairpin-shape. This intensifies the ejection, leading to further lifting and stretching. One structure can produce secondary structures up- and downstream through vortex roll-up of the ejected or sweeped fluid, respectively. This autogeneration process was confirmed by DNS studies (Zhou et al., 1999). The streamwise alignment of hairpins leads to zones of uniform momentum, i.e. elongated regions of $u' < 0$, which form a packet of structures which evolve and grow through the boundary layer as their are advected downstream. Further secondary structures are created in the wall region through roll-up of the downwashed fluid outside the wall-attached hairpin-legs (Brooke and Hanratty, 1993).

Adrian, Meinhart, and Tomkins (2000) summarize the length scales of hairpin vortices and packets as derived from particle-image-velocimetry for flows with different low Reynolds numbers: Near-wall hairpin structures have an approximate streamwise length of 200 viscous wall units δ_v with a distance of 50 δ_v between their legs, and the vortex heads are lifted at an angle of about 45° from the wall. They occur at a distance of several hundred δ_v in streamwise and about 100 δ_v in spanwise direction. The packets grow at a mean angle of 12° from the wall throughout the boundary layer. They can reach lengths of up to 2 z_i, which is also the length of the streaks induced by the packets' zones of uniform momentum.

Even though the descriptive model integrates earlier findings, it allows neither quantitative conclusions nor does it explain the complete spanwise coverage and regularity of the structures. Additionally, it is restricted to shear-flow without the influence of buoyancy.
An interesting aspect of the physical model is its contradiction of the common knowledge about the energy cascade in boundary layer flows, in which turbulent kinetic energy is transferred from larger to smaller eddies (cf. Chap. 2.1). Therefore, the bottom-up mechanism of the structure-packet evolution cannot be the only turbulence-generating mechanism (Adrian, 2007). Hunt and Morrison (2000) propose that, while dominant for small Reynolds numbers, the bottom-up mechanism is no longer valid for very high Re, where streaky structures are an effect of larger eddies impinging on the ground.

Young et al. (2002) notes two major competing mathematical theories for the creation and maintenance of streaks. According to Hamilton et al. (1995), streaks are created from streamwise vortices, which strengthen and become unstable, creating new vortices in their decay process. On the other hand, Foster (1997) showed that transient non-normal mode optimal perturbations can occur, which agree with the time and spatial scales of streaks. None of the theories has been verified or falsified yet (Young et al., 2002).

2.4. Detection and Characterization of Coherent Structures in the Atmospheric Boundary Layer

Atmospheric boundary layers flows are characterized by high Reynolds numbers and, apart from neutral stratification, by an important influence of buoyancy in the turbulence production.

The Reynolds number is a measure for the relation between the largest and smallest scales of turbulence (Eq. 2.6), therefore the wall region $(z/\delta_v \leq 100)$ takes up a considerably smaller portion in high-Reynolds number flows compared to smaller Re. As an example, $u_* = 0.3$ m/s and the kinematic viscosity of air, $v = 1.5 \cdot 10^{-5}$ m^2/s, result in $\delta_v = 5 \cdot 10^{-5}$ m, so the wall region only covers the lowest half centimeter of the boundary layer.

It is still unclear if the structure generation mechanisms of the hairpin-packet model are valid in the shear-driven atmospheric boundary layer (Adrian, 2007; Lin et al., 1996).

In the atmospheric boundary layer, three types of structures have been reported (Agee, 1984; Young et al., 2002): streaks and local vortical motions comparable to the hairpin vortices in shear-driven boundary layers (Hommema and Adrian, 2003; Newsom et al., 2008), horizontal convective rolls in moderately convective situations with shear (Etling and Brown, 1993; Hartmann et al., 1997), and polygonal spoke patterns in buoyancy-driven boundary layers without shear (Feingold et al., 2010).

The largest obstacle in atmospheric turbulence research is the long range of turbulent scales, which exacerbates measurements as well as simulations: no meteorological instrument captures all scales between the Kolmogorov microscale ($\mathcal{O}(1\ \mathrm{mm})$) and the largest scales ($\mathcal{O}(z_i)$). Likewise, the number of grid points required renders DNS modeling impractical with the currently available computational powers. Large-eddy simulations provide an intermediate solution: the turbulence is resolved down to a filter scale on the order of meters or tens of meters, and the smallest turbulence scales are parameterized (cf. Chaps. 3.2.1 and 3.2.2).

Scientific interest in the atmospheric boundary layer structures is founded in their contribution to the Reynolds stress tensor. In weather forecast models, the Reynolds stress appears as a variable which has to be parameterized using closure techniques. Up to now, organized motions are not considered in the sub-filter-scale parameterizations, which usually assume homogeneous and isotropic turbulence on the smaller scales (Doms et al., 2011). It can be expected that their influence becomes apparent when the forecast model grid resolution approaches typical structure length scales in the atmosphere. A parameterization which includes structures could therefore enhance mesoscale numerical models. Technically, this could be reached through a triple-decompositions of the flow-fields into the mean flow, the organized turbulence represented by the coherent structures, and the random turbulence (Hussain, 1983; Lykossov and Wamser, 1995; Hellsten and Zilitinkevich, 2013). The inhomogeneity apparent in the structured wind fields is also a candidate to explain the energy-balance closure problem, since it was shown that spatially averaged heat flux measurements yield higher fluxes than those detected with the eddy covariance method (Foken, 2008).

Apart from the phenomenology, understanding the formation and evolution of these surface layer structures could enhance insight in turbulent processes in general, including the initiation of convection.

2.4.1. Large-Eddy Simulations

Throughout the last years, several LES studies investigated coherent structures in boundary layers for varying magnitudes of shear and buoyant forcings.

Moeng and Sullivan (1994) and Lin et al. (1996) found the streaky ejection-sweep patterns in neutrally stratified boundary layers ($w_* = 0$). The anisotropy becomes less pronounced and the streaks become broader and fewer farther away from the surface. Using ensemble averages of strong ejections, Lin et al. (1996) showed that these conditional eddies in the surface layer are elongated in the mean-wind direction, with length scales of about 0.2 z_i, and become more circular in the mixed layer. The investigation of Khanna and Brasseur (1998) of horizontal integral length scales in the vertical wind component showed that the aspect ratio between the streamwise and spanwise component is $L_{wx}/L_{wy} \approx 9 - 12$ at the top of the surface layer and decreases linearly with z/z_i down to unity at the top of the boundary layer. Tracing the structures in time, Lin et al. (1996) showed that most eddies are generated in the surface layer and move upwards, always aligned with the local mean wind. During the process, most conditional eddies decay and only some reach and traverse the mixed layer. Remarkably, isosurfaces of vorticity around the conditional eddies resemble the hairpin-shapes found in low-Reynolds number simulations (Adrian, 2007), albeit at a larger scale. Lin et al. (1997) derived an equation for the streak-spacing λ in the neutral boundary layer:

$$\frac{z}{z_i} = a + b \cdot \frac{\lambda}{z_i} , \qquad [2.7]$$

with $a = -0.24 \pm 2.3 \cdot 10^{-2}$ and $b = 0.56 \pm 3.38 \cdot 10^{-2}$.

Results from Khanna and Brasseur (1998) confirm the linear increase from $\lambda/z_i \approx 0.5$ close to the ground for stabilities $-z_i/L_* \leq 8$, with b becoming smaller for larger $-z_i/L_*$.

For increasingly convective situations, the surface layer structures are tilted away from the ground (Kim and Park, 2003) and they begin to extend higher into the boundary layer (Khanna and Brasseur, 1998). Kim and Park (2003) showed for $-z_i/L_* = 1.95$ that ejections contribute 75% to the upward momentum flux at the bottom of the mixed layer.

In the stability regime $1.5 \leq -z_i/L_* \leq 9.5$ some streaks develop into horizontal rolls which reach from the surface to the top of the boundary layer (Sykes and Henn, 1989; Moeng and Sullivan, 1994) with a spanwise spacing of 2 to 3 z_i (Khanna and Brasseur, 1998; Moeng and Sullivan, 1994). An example from LES is shown in Fig. 2.3. Meanwhile, the surface layer remains populated with low-rise streaks, the vertical growth of which is suppressed by the roll-induced downdrafts. The aspect ratio between streamwise and spanwise scales remains unchanged by the onset of roll convection (Khanna and Brasseur, 1998).

The horizontal integral length scales studied by Khanna and Brasseur (1998) showed an increase in structure length scale up to $z/z_i \leq 0.5$ and a subsequent decrease for both the convective and the shear-dominated simulations. The aspect ratio between streamwise and spanwise integral scales was found to be $L_{wx}/L_{wy} \leq 2$ for $-z_i/L_* \geq 10$.

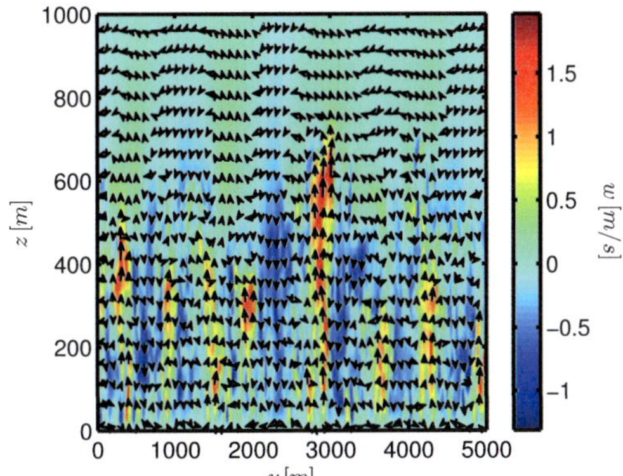

Figure 2.3: Vertical wind field (color) in a spanwise vertical plane from LES with $-z_i/L_* = 7.2$ (cf. Tab. 5.2). The vectors show the projection of the wind field on the plane.

For buoyancy-driven boundary layers with very small u_*, Khanna and Brasseur (1998, with $u_* = 0.16, -z_i/L_* = 841$) find that hexagonal structures develop with narrow updraft regions enclosing larger downdraft-cells. Those structures are associated with Rayleigh-Bénard cell convection (Lord Rayleigh, 1916). Hellsten and Zilitinkevich (2013) find that these structures contribute more than 90% to the momentum flux in the mixed layer. The cells disappear quickly as soon as the shear gains influence (Moeng and Sullivan, 1994, with $u_* = 0.56, -z_i/L_* = 19$). These structures show a similar behavior to the convective streaks and rolls, i.e. the updraft regions become broader and less intense with height, summarizing smaller-scale structures from below and suppressing low-rise structures in the downdraft regions.

Although LES have proven to be reliable boundary layer models, small-scale turbulence close to the wall cannot be resolved, which consequently makes the properties of surface layer structures sensitive to the subgrid-scale model (Khanna and Brasseur, 1998). Additionally, the lowest grid-level is often parameterized using Monin-Obukhov similarity (Raasch and Etling, 1991). However, shear-generated vortical structures appear to have their source near the surface before they grow through or traverse the boundary layer. It is therefore necessary to supplement the LES results with measurements to validate the model results.

2.4.2. Atmospheric Boundary Layer Observations

Large-scale convective structures can become apparent in the cloud structure on top of the boundary layer. Cloud streets of several kilometer length have been observed atop cold-air outbreaks over oceans, indicating the formation of counter-rotating horizontal convective rolls (Hartmann et al., 1997; Brümmer, 1999). However, Etling and Brown (1993) note that the observation of clouds alone is not sufficient to quantify the roll scales and spanwise spacing.

Hexagonal patterns have been observed in the cloud tops of convection driven boundary layers (Feingold et al., 2010). Here, open cells form with narrow updraft bands enclosing larger downdraft areas when the convection is driven by heating from the bottom boundary, whereas cooling at the top leads to a closed cell structure with narrow downdrafts enclosing larger updraft areas.

Hairpin-like surface layer structures and packets were visualized by Hommema and Adrian (2003) using smoke as a passive marker over a desert floor in the nighttime. The observed packet growth angles agreed with the hairpin-packet model of Adrian, Meinhart, and Tomkins (2000).

As discussed above, quantitative atmospheric turbulence research faces the challenge of having to simultaneously capture data from a large volume of air with high time and spatial resolution.

Traditionally, meteorological towers were used for high-resolution point measurements while assuming Taylor's hypothesis of frozen fields to infer the spatial structure (Stull, 1988). In this manner, the streamwise wind field can be investigated. The most notable coherent structure detection technique used on tower time series of wind field and temperature data in recent years is the wavelet analysis (Collineau and Brunet, 1993a). With this technique the expected ejection-sweep-patterns can be detected, and their contribution to TKE and turbulent fluxes on different length scales can be determined.

The method has been extensively used in recent years on time series from tower data at various heights in the surface layer (Lykossov and Wamser, 1995), and especially to investigate the flow structure in and atop forest canopies (Collineau and Brunet, 1993a,b; Thomas and Foken, 2007; Segalini and Alfredsson, 2012; Zeeman et al., 2013). Depending on stability, measurement and canopy height and the particular detection technique the length scales of structures vary between few tens of meters and almost one kilometer (Barthlott et al., 2007). Likewise, their relative contribution to the turbulent fluxes is determined to lie between 10% (Zhang et al., 2011) and 100% (Feigenwinter and Vogt, 2005). Barthlott et al. (2007) give a summary over the studies before 2007.

In general, the structures become more elongated as the stratification becomes more unstable, and shorter again for very unstable situations (Thomas and Foken, 2005; Barthlott et al., 2007). The characteristic ejection-sweep-patterns are observed in the time series of the wind components u' and w', as well as those of temperature T' and humidity q'. For u' and w', the structure intensity is approximately proportional to u_* (Zhang et al., 2011). Attempts have been made to retrieve the structure shape in the x-z-plane from simultaneous measurements at different heights, but the spanwise component remains unretrievable from tower data.

Lenschow and Stankov (1986) studied the horizontal autocorrelation of the wind field in the convective boundary layer ($10 \leq -z_i/L_* \leq 62$) with aircraft measurements, and found for the mean streamwise horizontal integral length scales L_α for the wind component α, that

$$\frac{L_w}{z_i} = 0.28 \left(\frac{z}{z_i}\right)^{1/2}, \qquad\qquad [2.8a]$$

$$\frac{L_u + L_v}{2z_i} = 0.53 \left(\frac{z}{z_i}\right)^{1/2}. \qquad\qquad [2.8b]$$

The results reflect the structure growth through the boundary layer and agree well for the vertical wind at $z/z_i \leq 0.5$ in the two most convective cases of Khanna and Brasseur (1998), and even approximately for the shear-dominated case. The aspect ratio between streamwise and spanwise integral scales was found to be $L_{wx}/L_{wy} \leq 2$ for $-z_i/L_* \geq 10$.

Inagaki and Kanda (2010) used 40 sonic anemometers to characterize the surface-layer flow and were able to visualize the streaky structures. However, the setup is rather impractical for high-resolution measurements of large-scale structures.

To overcome these issues, remote sensing instruments have increasingly been used for atmospheric flow measurements, since they provide a high time and spatial resolution over long ranges. Kropfli and Kohn (1978) detected horizontal convective rolls with Doppler radar. The smaller-scale surface-layer streaks are best investigated using high-resolution Doppler lidars. Drobinski et al. (1998) furthermore used a Doppler lidar and a sodar to investigate horizontal convective rolls.

When deployed in dual-lidar mode, the complete horizontal wind field can be retrieved in an area of several square kilometers. Newsom et al. (2008) used this method to detect surface layer streaks and measure the streamwise and spanwise correlation length of the wind field. They found that the integral length scales of the streamwise wind component became maximal for neutral conditions, with $L_x \approx z_i$, and $L_x \approx 0.5 z_i$ for weakly stable and unstable conditions.

Tang et al. (2011a,b) analyzed the horizontal wind field from lidar measurements in terms of Lagrangian coherent structures (Shadden et al., 2005) to detect vertical gusts and even footprints of hairpin structures on airport runways. Lagrangian coherent structures are persistent barriers in two-dimensional flow derived from the flow-field trajectories and which are observed frequently in ocean currents (Lekien et al., 2005). Close to the atmospheric boundary layer surface, flow barriers coincide with updrafts, so Lagrangian coherent structures can be indicative of positive vertical wind velocities and thereby used to measure convective cell patterns.

Dual-lidar deployments can also be used for volume measurements as in Iwai et al. (2008), who showed that the surface-layer streaks are the starting points for horizontal convective rolls with a spanwise spacing of approximately 2 z_i.

As a goal, the research of boundary layer coherent structures should lead to a parameterization of their scales and contribution to the Reynolds stress to improve sub-filter-scale parameterizations in mesoscale models and thereby enhance their accuracy. The evidence shows that their length scales, spacings and intensity are influenced by shear, stability, and boundary layer height. The surface roughness length may play a role as well (Lin et al., 1997). However, no consistent parameterization is available up to now.

The focus lies here on the detection of recurring patterns in the flow field, without too much emphasis on the question of what exactly constitutes a coherent structure. The exact separation of the flow field in a structured and an unstructured part should be motivated by practical considerations concerning parameterization, respective contributions to the fluxes, and their coupling terms. This can only take place after a reliable detection method is established.

3. Dual-Doppler Lidar: Measurements and Simulations for Coherent Structure Detection

In this chapter, the lidar measurement principle is introduced and, based on its mathematical description, a pulsed Doppler lidar simulation tool is developed which operates on LES simulations with the PALM model. Subsequently, a retrieval algorithm is discussed which allows to re-assemble the two-dimensional wind field from either virtual or real dual lidar planar scans.

3.1. Pulsed Doppler Lidar Measurements

3.1.1. Measurement Principle of Pulsed Doppler Lidars

The acronym *lidar* stands for LIght Detection And Ranging and was created based on the word *radar* (RAdiowave Detection And Ranging, Middleton and Spilhaus, 1953). A lidar emits laser radiation into the atmosphere and detects the scattered return signals. For pulsed lidars, the time lapse between laser emission and detection of the scattered light can be used to determine the position of the scatterers along the lidar beam.

A laser consists in general of an active medium, an energy pump and an optical resonator (Demtröder, 2009). In thermal equilibrium, the energy states E_k and E_i with $E_k < E_i$ in the active medium have a population $N_k > N_i$. The pump is used to create a population inversion, i.e. it stimulates transitions into the higher state E_i, until N_i is large enough compared to N_k that an incoming photon $h\nu = E_i - E_k$ will not lead to an excitation $E_k \rightarrow E_i$, but rather lead to a stimulated emission of another photon: $E_i \rightarrow E_k$. The multiplication of photons, reflected back and forth in the resonator, leads to a cascade of stimulated emissions: a pulse of high-intensity, coherent, monochromatic light.

The power of the backscattered lidar return signal P is related to the power of the outgoing laser pulse P_0 via the lidar equation (Klett, 1981):

$$P(r,\lambda) = P_0 \frac{c\tau}{2} \frac{A}{r^2} \beta(r,\lambda) e^{-2\int_0^r dr'\, \alpha(r',\lambda)} , \qquad [3.1]$$

where r is the signal origin along the beam, λ the pulse wavelength, τ the temporal pulse width, A the detector area, β the backscatter coefficient related to the scatterer concentration and their scattering cross section, and α the atmospheric extinction coefficient.

Common types of lidar are (Wandinger, 2005):

- The elastic-backscatter lidar, which measures properties of aerosols and clouds from their elastic scattering properties in the return signal (Spuler and Mayor, 2005),

- The differential-absorption lidar (DIAL), which is used to measure the concentration of atmospheric trace gases, e.g. ozone and water vapor, from their different absorption coefficients for different wavelengths (Wulfmeyer and Bösenberg, 1998),

- The Raman lidar, which detects gases, especially water vapor, from the Raman scattering return signals and can be used for temperature profiles (Radlach et al., 2008),

- The resonance scattering lidar, which detects molecules and ions from resonant fluorescent scattering at known energy transitions (Alpers et al., 2004),

- The Doppler lidar, which is used to measure the velocity of aerosols and molecules from the Doppler shift in the return signal.

This work is focus pulsed coherent or heterodyne-detection Doppler lidars. The heterodyne technique mixes the monochromatic lidar pulses with frequency f_0, which are emitted into the atmosphere, and the return signal with the Doppler shift Δf with the signal of a local oscillator (LO) of known frequency f_{LO}. The intensity of the resulting signal is given by (Werner, 2005)

$$I \propto \cos(2\pi\,[f_{LO} - (\Delta f + f_0)]) + \cos(2\pi\,[f_{LO} + (\Delta f + f_0)]) \,. \qquad [3.2]$$

The high-frequency part of the signal is filtered out, whereas the first part, the so-called beat signal, has a low frequency which can be analyzed with high accuracy using a Fast Fourier Transform (FFT).
The Doppler shift

$$\Delta f = -f_0 \cdot 2\frac{v_r}{c} \,, \qquad [3.3]$$

where v_r is the local wind vector projected on the lidar beam direction (the so-called radial or line-of-sight wind speed) and c is the speed of light (Werner, 2005). For $v = 1$ m/s this results in a shift of only 1 MHz for $f_0 = 1.5 \cdot 10^{14}$ Hz.
The lidars usually operate with frequencies f_0 in the infrared for which Mie-scattering by aerosols exceeds Rayleigh-scattering by air

molecules. All particles exhibit random motion with a kinetic energy proportional to their temperature superimposed on their mean motion, but since aerosols have a higher mass their velocity fluctuations are smaller, which in turn leads to less spectral broadening in the return signal (Werner, 2005). The Doppler lidar therefore detects the radial velocity of aerosols, which is assumed to equal the wind velocity.

Pulsed Doppler lidars emit laser pulses at a given pulse repetition frequency (PRF) and record the return signals with a certain sampling rate (SR). The number of samples (SpG) used for the FFT corresponds to a segment of the lidar beam, the so-called range gate, with length

$$\Delta p = \frac{\text{SpG} \cdot c}{2 \cdot \text{SR}} \, . \tag{3.4}$$

For some instruments, the user can define the lengths of the range gates and their distribution along the beam. The detected Doppler shift is a weighted average produced by the radial velocities of all particles illuminated by the pulse moving through the range gate. An estimate of the measured radial velocity $rv(R_0)$ for a range gate centered around R_0 is usually computed from the average spectra of several consecutive pulses to decrease the random error (Frehlich, 1997).

A mathematical model for solid-state pulsed Doppler lidar velocity estimation was given by Frehlich et al. (1998):

$$rv(R_0,t) = \int_{-\infty}^{\infty} \mathrm{d}x\, v_r(x,t) W_{\Delta p}(x - R_0) \, , \tag{3.5}$$

with the normalized weighting function

$$W_{\Delta p}(x) = \int_{-\infty}^{\infty} \mathrm{d}r\, I_n(x - r)\theta_{\Delta p}(r) \, , \tag{3.6}$$

defined by a range gate indicator function $\theta_{\Delta p}$,

$$\theta_{\Delta p}(x) = \begin{cases} 1/\Delta p & , x \in [-\Delta p/2, \Delta p/2] \\ 0 & , \text{otherwise} \end{cases} \quad , \qquad [3.7]$$

which is unity on the range gate and zero otherwise, and the Gaussian pulse envelope of the beam:

$$I_n(x) = \frac{2}{\sqrt{\pi}\sigma_\tau c} e^{-\frac{4x^2}{\sigma_\tau^2 c^2}} \quad , \qquad [3.8]$$

where σ_τ is the standard deviation of the pulse in time domain.

Depending on the velocity estimator of the lidar system, it is also possible to have a tapered gate window and thereby a non-uniform range gate indicator function $\theta_{\Delta p}$ (Kristensen et al., 2010).

Eqs. 3.6 - 3.8 show that the weighting function is only a function of Δp, σ_τ and the distance x from the range gate center, therefore according to similarity theory (Buckingham, 1914) the dimensionless weighting function $W_{\Delta p} \cdot \Delta p$ can be written as a function of the dimensionless groups $x/\Delta p$ and $\Delta p/(\sigma_\tau c)$:

$$W_{\Delta p}(x) \cdot \Delta p = \tilde{W}\left(\frac{x}{\Delta p}, \frac{\Delta p}{\sigma_\tau c}\right) \quad , \qquad [3.9a]$$

$$\text{with} \qquad \tilde{W}(a,b) = \int_{-1/2}^{1/2} da' \frac{2b}{\sqrt{\pi}} e^{-4b^2(a-a')^2} \quad . \qquad [3.9b]$$

Consequently, the relative weight at a distance $x/\Delta p$ from the range gate center only depends on the relation of range gate length to pulse with, which is shown in Fig. 3.1: For $\Delta p \gg \sigma_\tau c$, the weighting function approaches the range gate indicator function $\theta_{\Delta p}$ (Eq. 3.7), whereas for $\Delta p \ll \sigma_\tau c$ it approaches the pulse envelope I_n (Eq. 3.8). The lidar resolution is therefore naturally limited by the pulse width. For a laser pulse

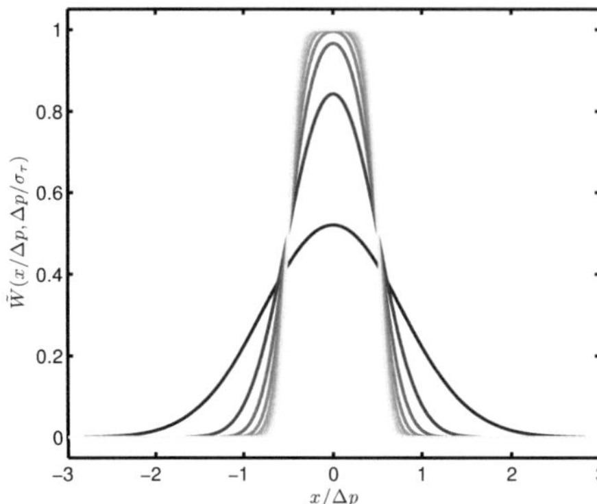

Figure 3.1: The lidar weighting function in relative coordinates according to Eqs. 3.9 as a function of $\frac{x}{\Delta p}$ for $\frac{\Delta p}{\sigma_\tau c} = \{0.5, 1, 1.5, 2, 2.5, 3, 3.5, 4\}$ (darker shades mean smaller values).

to be considered monochromatic its width in the frequency domain must be very small. Since the width of a Gaussian in frequency and time domain are inversely proportional, σ_τ cannot become smaller without a loss in wavelength accuracy. This inherent spatial averaging means that Doppler lidars can never perform point measurements.

For scanning or moving lidars, producing a velocity estimate from several consecutive pulses means that the beam movement must be included in the weighting function. This can be accomplished with a linear average in the direction of the beam movement (Frehlich, 2001):

$$rv(R_0,t) = \int\limits_{y_1}^{y_2} \mathrm{d}y \int\limits_{-\infty}^{\infty} \mathrm{d}x\, v_r(x,y,t)\, W_{\Delta p}(x - R_0) , \qquad [3.10]$$

when during one velocity estimate the range gate center moves from y_1 to y_2 on the y-axis, which is defined by the range gate center trajectory. The resulting relative weights for a scanning beam are shown in Fig. 3.6.

In recent years, long-range high-resolution Doppler lidars have become increasingly stable and affordable. As an ideal remote-sensing instrument for boundary layer wind-field research, they have been used to study the convective boundary layer (Lothon et al., 2006), convective roll vortices (Drobinski et al., 1998), entrainment processes on the top of the mixed layer (Träumner et al., 2011), the nocturnal low-level jet (Banta et al., 2002) and the cloud-topped boundary layer (Lottman et al., 2001). They also play a role in engineering application, measuring wake vortices behind wind turbines (Krishnamurthy et al., 2013; Käsler et al., 2010). Their deployment in dual- or even multi-Doppler mode is discussed in Chap. 3.1.3.

3.1.2. The KIT Doppler-Lidar Systems

At Karlsruhe Institute of Technology (KIT), the Institute for Meteorology and Climate Research (IMK-TRO) operates two coherent pulsed Doppler lidars of the 'WindTracer'-type. The instruments were manufactured by Lockheed Martin Coherent Technologies, Inc.(LMCT), Louisville, Colorado, USA.

The sensitive systems are encased in containers with heat and humidity control, which in turn are mounted on swap body structures for easy transportation with trailers. The instruments have been used in several international measurements campaigns, e.g., the Convective and Orographically-induced Precipitation Study (COPS, Kottmeier et al., 2008; Wulfmeyer et al., 2008), the HYdrolocial cycle in the Mediterranean EXperiment (HyMeX, Kalthoff et al., 2013), and the HOPE experiment as a part of HD(CP)^2 (High Definition Clouds and Precipidation for advancing Climate Prediction).

Tab. 3.1 gives an overview of the technical specifications of the lidar systems. Both systems have solid-state lasers (thulium-doped lutetium

'WindTracer'	1	2
year of construction	2004	2009
type of laser	Tm:LuAG	Er:YAG
wavelength	2023 nm	1617 nm
pulse length	370 ns	300 ns
pulse energy	2.0 mJ	2.7 mJ
pulse repetition frequency	500 Hz	750 Hz
sampling rate	250 MHz	250 MHz

Table 3.1.: Technical specification of the KIT 'WindTracer' systems.

aluminum garnet in WindTracer 1, and erbium-doped yttrium aluminum garnet in WindTracer 2). The emitted pulses conform to the Gaussian approximation (Eq. 3.8) with a standard deviation of $\sigma_\tau = 370$ ns and 300 ns, which correspond to a full width at half maximum (FWHM) of 92 m and 75 m, respectively (Frehlich et al., 1998). The high pulse repetition frequency (500 Hz and 750 Hz) facilitates highly accurate radial velocity estimations with a measurement frequency of up to 10 Hz. With wavelengths larger than 1.4 μm, the lasers are considered eye-safe (Henderson et al., 1993).

The instruments have a range of up to 12 km under clear conditions and a pulse width of approximately 70-90 m. An almost identically constructed lidar system was described in detail by Grund et al. (2001).

Because of their equality, the systems can be operated in dual-Doppler mode, i.e. they can be steered synchronously in coordinated measurements.

A common challenge in many dual-Doppler measurements is the time-synchronization (Calhoun et al., 2006): for scanning lidars, a high time-resolution in the measurements requires an agreement of the systems clocks over the duration of the measurement, as well as scanning patterns which do not accumulate relative phases shifts. This issue commonly prevents measurements from being conducted with the highest achievable time resolution (cf. Newsom et al., 2008). Often this problem arises from restrictions in the beam-steering software: many lidar systems only allow pre-defined PPI (plan-position indicator, e.g. fixed elevation angles el) and RHI (range-height indicator, i.e. fixed azimuth angles az) scans, but no free beam-steering (cf., e.g., Grund et al., 2001).

The KIT dual-Doppler lidar system can be operated by a unique control software, which was developed as a part of this work. This software runs on an external PC, the Remote Operating Station (ROS), which is connected to both lidars and receives their status updates every second. It is based on a C-library of basic control functions for the single lidar systems supplied by the manufacturer, which can be used to set the lidar control parameters and steer the beams. By combining the single lidar controls in dual-lidar steering functions in a C++-based library, it became possible to program complex scanning patterns which synchronize automatically without relying on the single lidar clocks. Important functions in the control software are:

- Setting lidar control parameters like range gate length, measurement frequency, position of range gate centers, recorded data types and others,

- Beams steering from one (az,el)-position to another with a constant angular velocity for each angle,

- Dual-lidar synchronization by including waiting intervals until each lidar is at the desired position,

- Real-time adaptability of scan patterns to external parameters, line wind direction and boundary layer height.

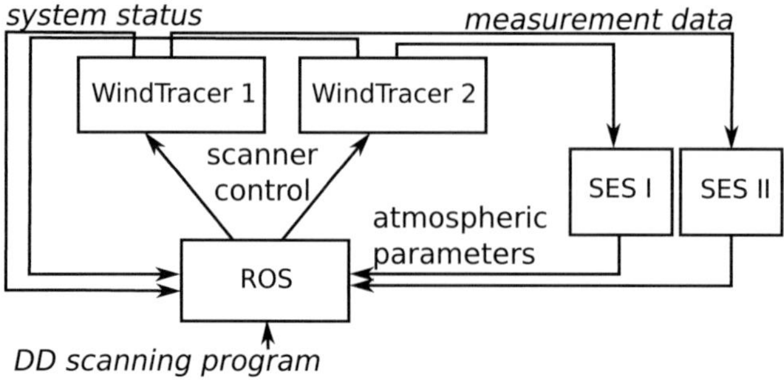

Figure 3.2.: Stawiarski et al. (2013): Schematic overview of the dual-Doppler control system. 'Wind-Tracer 1' and 'WindTracer 2' are the two control computers of the individual instruments, ROS is the remote operation station, SES are storage and evaluation stations. ©2013 American Meteorological Society. Used with permission.

With this software, repetitive complex scanning patterns can easily be realized, e.g. the synchronized scan of a plane alternating with RHI or velocity-azimuth-display scans for vertical profiles (VAD, cf. Browning and Wexler, 1968).

Each lidar stores its measurement data on a Storage-and-Evaluation Station (SES). On these computers, MATLAB-based programs evaluate the data in real time: vertical wind profiles (from VAD and RHI scans) and the boundary layer height (from vertical stares) are computed whenever suitable scans were performed. The SES send the results to the ROS, thus enabling users to program scans which adapt to the current atmospheric conditions. Possible applications for this feedback-loop are coplanar-scan optimization with respect to the horizontal wind speed (Chap. 4.2) and virtual tower measurements where the tower height adapts to the boundary layer height (Röhner and Träumner, 2013). The local network is sketched in Fig. 3.2.

3.1.3. Dual-Doppler Measurements

Operating two lidars in dual-Doppler mode means that velocity estimates are obtained simultaneously from two lidars at (approximately) the same point in space. Thereby two linearly independent components of the wind fields are measured as long as the beams are not parallel, from which the two-dimensional wind vector at the crossing point in the lidar plane (i.e., the plane spanned by the beams) can be deduced.

Two examples for dual-lidar set-ups are the intersecting beams technique, which is used e.g. in virtual tower measurements, and the planar scan technique.
In the former, the wind field is only retrieved on the trajectory of the beam intersection, which yields a point measurement with a high frequency. During the planar scan technique, both lidars scan the same area to obtain the wind field in the full overlap region, albeit with a decreased time resolution.

Dual-Doppler Intersecting Beam Techniques

Assume two lidars with current azimuth and elevation angles $\{az_i, el_i\}$, $i = 1, 2$. The unit vectors of their beam directions in Cartesian coordinates are then given by

$$\hat{\mathbf{r}}_i = \begin{pmatrix} \sin(az_i)\cos(el_i) \\ \cos(az_i)\cos(el_i) \\ \sin(el_i) \end{pmatrix} , \, i = 1, 2 \quad . \qquad [3.11]$$

If the beams intersect at a point $\mathbf{x}$, at each point in time the velocity estimates $\{rv_1, rv_2\}$ measured by the lidars in the range gates closest to $\mathbf{x}$ can be used to derive the wind vector:

$$\begin{pmatrix} rv_1(t) \\ rv_2(t) \end{pmatrix} = \begin{pmatrix} \hat{\mathbf{r}}_1^T \\ \hat{\mathbf{r}}_2^T \end{pmatrix} \cdot \mathbf{u}(\mathbf{x},t) \, . \qquad [3.12]$$

The system of equations 3.12 is underdetermined for a wind vector with three components. In long-time averages, it is often assumed that the vertical wind component is zero, thereby reducing the system to two equations and rendering it solvable (cf. Calhoun et al., 2006). However, to reap the full advantage of the high time resolution, no such assumption can be made in the turbulent boundary layer, and only the two-dimensional projection $\mathbf{u}_H$ of the wind vector $\mathbf{u}$ on the lidar plane can be retrieved:

$$\mathbf{u}_H = \mathbf{u} - \left(\mathbf{u} \cdot \hat{\mathbf{n}}_n \right) \hat{\mathbf{n}}_n \, , \qquad [3.13]$$

where $\hat{\mathbf{n}}_n = \hat{\mathbf{r}}_1 \times \hat{\mathbf{r}}_2 / \parallel \hat{\mathbf{r}}_1 \times \hat{\mathbf{r}}_2 \parallel$ is the normal vector of the plane spanned by the lidar beams.

Usually, a local coordinate system is defined in the lidar plane. The component u_j of $\mathbf{u}_H$ on the axis determined by the direction of the normalized vector $\hat{\mathbf{e}}_j$ can be derived from the unique linear combination of the $\hat{\mathbf{r}}_i$ which forms $\hat{\mathbf{e}}_j$:

$$\hat{\mathbf{e}}_j = q_1 \hat{\mathbf{r}}_1 + q_2 \hat{\mathbf{r}}_2 \qquad [3.14a]$$

$$\Rightarrow u_j = \mathbf{u} \cdot \hat{\mathbf{e}}_j = q_1 \, rv_1 + q_2 \, rv_2 \qquad [3.14b]$$

Note that $\mathbf{u} \cdot \hat{\mathbf{e}}_j = \mathbf{u}_H \cdot \hat{\mathbf{e}}_j$, since $\hat{\mathbf{e}}_j \perp \hat{\mathbf{n}}_n$.

Fig. 3.3 shows the lidar plane. The intersecting beam angle $\Delta\chi$, the angle α_j between $\hat{\mathbf{e}}_j$ and the mean lidar beam direction $\mathbf{r}_m$, and the angle γ_{u_H} between the wind vector $\mathbf{u}_H$ and the direction of evaluation $\hat{\mathbf{e}}_j$ determine the relative position of these four vectors. The pre-factors q_1, q_2 in

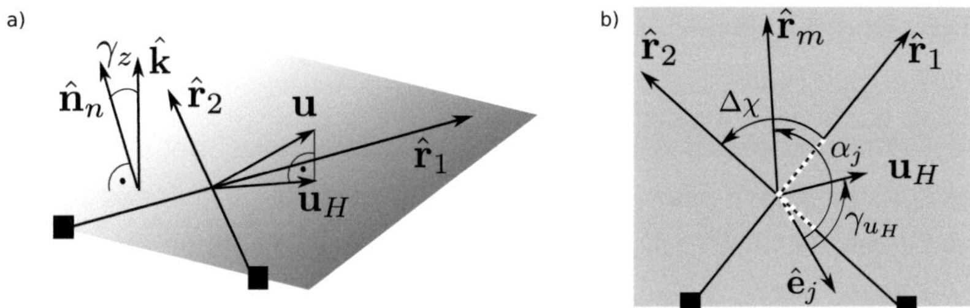

Figure 3.3.: Stawiarski et al. (2013): Relevant vectors and angles in the lidar plane. The squares denote the lidar positions. a) Lidar beam vectors $\hat{\mathbf{r}}_1, \hat{\mathbf{r}}_2$ spanning the lidar plane (shaded area). $\hat{\mathbf{n}}_n$ is the plane normal vector, $\hat{\mathbf{k}}$ points in the vertical direction. The plane is tilted away from the horizontal by an angle γ_z. The wind vector $\mathbf{u}$ is projected on the plane to give $\mathbf{u}_H$, the retrievable wind vector. b) View on the lidar plane with beams $\hat{\mathbf{r}}_1, \hat{\mathbf{r}}_2$, planar wind vector $\mathbf{u}_H$, direction of evaluation $\hat{\mathbf{e}}_j$ and mean lidar beam direction $\hat{\mathbf{r}}_m$. The three angles suffice to fix the relative vector positions. All angles are measured in the positive sense, i.e. counter-clockwise. The dotted lines indicate the projections of $\mathbf{u}_H$ on $\hat{\mathbf{r}}_1, \hat{\mathbf{r}}_2$ and $\hat{\mathbf{e}}_j$ with lengths $rv_1, -rv_2$ and u_j, respectively (note that $rv_2 < 0$). ©2013 American Meteorological Society. Used with permission.

Eq. 3.14b can be expressed in these angles, which yields (Stawiarski et al., 2013):

$$u_j = \mathbf{u} \cdot \hat{\mathbf{e}}_j = \frac{rv_1 \sin(\alpha_j + \frac{\Delta\chi}{2}) - rv_2 \sin(\alpha_j - \frac{\Delta\chi}{2})}{\sin(\Delta\chi)} . \qquad [3.15]$$

The intersecting beam technique with a high time resolution in the lidar plane was used by Collier et al. (2005) in the Invest-to-Save Budget project 52 (ISB52) for the validation of dispersion models in the boundary layer. In this measurement, a time resolution of 5 s was achieved.

On the other hand, Calhoun et al. (2006) used the technique for longer averaging times, so that the assumption $\overline{w} = 0$ became valid. Furthermore, the averaging allowed them to scan along the height range of virtual towers without synchronization, since the time shifts become irrelevant compared to the averaging interval. This method is referred to as the 'virtual tower' technique.

Dual-Doppler Planar Scan Techniques

In planar scanning patterns, the wind field is retrieved not only at the beam intersection point, but on a larger area which is scanned by the two lidars. Geometrically, any such plane is fixed by the position of the two lidars and a slope, which determines the azimuth and elevation angles during the scan. Except for horizontal and vertical planes, which can be realized with fixed elevation an azimuth angle scans, respectively, this requires a lidar steering software in which azimuth and elevation can both vary smoothly during a scan.

To retrieve the two-dimensional wind field in the lidar plane one has to assume that the variation of the radial wind velocity is negligible during the time of a beam sweep, since the lidar beams do not traverse each point of the plane at the same time. Consequently, the time resolution of the retrieved wind field is determined by the duration of the beam sweeps. The spatial resolution in the plane depends on the range gate length and lidar pulse width. However, to account for the uneven distribution of measurement points in the plane, the area is usually divided into grid cells which contain several velocity estimates, the weighted average of which is used to retrieve the velocity using Eq. 3.15. One possible retrieval algorithm is described in Chap. 3.3.

Planar scan patterns have been used by Newsom et al. (2008) and Iwai et al. (2008) (low-elevation sector PPI scans to retrieve the horizontal wind field), as well as Hill et al. (2010) (RHI scans to retrieve the vertical wind and the horizontal wind in the direction of the lidar connection line). Iwai et al. (2008) furthermore extended the measurement to higher elevations for a three-dimensional retrieval of u and v, the horizontal wind components.

Towards a retrieval of the three-dimensional wind field

In principle, the three-dimensional wind field can be retrieved from the velocity estimates of three Doppler lidars (Mann et al., 2009). However, research in this area is rare, which is probably due to the high acquisition and maintenance costs of Doppler lidars. Instead, attempts have been made to retrieve the three-dimensional wind field from single and dual lidar data. This requires further assumptions about the wind field. Less computationally expensive models retrieve volume data of the horizontal wind field from volume scans and deduce the vertical wind component from the integration of the continuity equation between horizontal layers (Drechsel et al., 2009; Iwai et al., 2008). Newsom et al. (2005) point out that this method has shortcomings for rapidly evolving structures in the fields. Therefore, a more complex four-dimensional variational data assimilation technique (4DVAR) is used often, where the output from single (Chai et al., 2004) and dual lidar data (Newsom et al., 2005; Xia et al., 2008) is fitted to a dynamical model to retrieve the complete wind field. The disadvantage lies here in high computational costs and the underlying assumptions in the model.

The general disadvantage of volume scans is the poor time resolution (e.g., 172 s for Lin et al., 2008). A higher resolution can be achieved with planar scans and a one-step integration of the continuity equation from the ground to the observation height, which is attempted in Chap. 7.

3.2. Simulations of Doppler-Lidar Measurements

The ability to retrieve horizontal wind fields with a resolution of the order of tens of meters is a unique feature of planar dual-Doppler lidar measurements. As a consequence, the whole data set cannot be compared with other instruments. The theory of Chap. 3.1.1 implies that the dual-lidar measurement acts like a low-pass filter in time and both spatial directions and on both components, with a spatial filter length given by the range gate length and/or the pulse width, although the exact filter function remains unknown.

A Doppler lidar simulator in combination with a realistic turbulence-resolving atmospheric model can help to transfer the theoretical knowledge about single lidar measurements to predictions about the performance of dual-lidar retrieval data in coherent structure detection schemes: Detection algorithms can be applied to the high-resolution boundary layer model data and to the simulated dual-lidar measurements in the model, and a comparison can be used to assess the agreement of results and potentially to correct the lidar results.

For realistic comparisons, it is essential that the lidar simulator produces velocity estimates which are in accordance with the theoretical model. Drechsel et al. (2010) developed a dual-lidar simulation scheme to optimize volume scans, which was based on analytical mesoscale model wind fields with 100 m resolution, in which the virtual lidar performed point measurements of the radial velocity while the field was kept stationary. The rapidly evolving small-scale structures of the surface layer however impose stronger demands on the time and spatial resolution. Therefore, a lidar simulator was developed (Chap. 3.2.3) which is applied to large-eddy simulations (Secs. 3.2.1, 3.2.2) with a grid spacing much smaller than the lidar averaging scale, which allows all lidar-relevant turbulent scales to be resolved in the model. The simulator is based solely on the mathematical description (Chap. 3.1.1), and thereby includes the important aspects of lidar measurements: the averaging in beam direction as determined by pulse shape and range gate length, the cross-beam linear averaging for scanning beams, and the full time resolution.

3.2.1. Large Eddy Simulations

The atmospheric boundary layer is a fluid with a high Reynolds number which exhibits turbulent flow. Conservation of momentum is described by the Navier-Stokes equations (e.g. Etling, 2008),

$$\frac{\partial u_i}{\partial t} = -u_j \frac{\partial u_i}{\partial x_j} - g\delta_{i3} - \varepsilon_{ijk}f_j u_k - \frac{1}{\rho}\frac{\partial p}{\partial x_i} + v\left(\frac{\partial^2 u_i}{\partial x_j^2} + \frac{1}{3}\frac{\partial^2 u_j}{\partial x_i \partial x_j}\right), \quad [3.16]$$

with the wind vector components u_i, $i = 1,2,3$, the pressure p, the density of air ρ, the molecular kinematic viscosity v, the coriolis parameter $\mathbf{f} = (0, 2\Omega\cos\varphi, 2\Omega\sin\varphi)$ at latitude φ and angular frequency Ω of the earth rotation, and the gravitational acceleration g. Summation over repeated indices is implied.

In combination with the conservation of mass, described by the continuity equation,

$$\frac{\partial \rho}{\partial t} + \frac{\partial(\rho u_j)}{\partial x_j} = 0, \quad [3.17]$$

the conservation of energy, which is the first law of thermodynamics,

$$\frac{\partial \theta}{\partial t} = -u_j \frac{\partial \theta}{\partial x_j} + v_\theta \frac{\partial^2 \theta}{\partial x_j^2} + Q_\theta, \quad [3.18]$$

with the potential temperature θ, thermal diffusivity v_θ and the source term Q_θ, and the ideal gas law,

$$p = \rho R_L T, \quad [3.19]$$

with the gas constant R_L of air, one obtains a set of six governing equations for the dry atmosphere (cf., e.g., Etling, 2008; Stull, 1988).

These are the equations of motion of the atmospheric state variables $\mathbf{u}$, p, ρ, and θ (the potential temperature). Moisture content or other scalar quantities can be included with further equations of conservation.

Atmospheric turbulence can be regarded as a superposition of interacting vortices ('eddies') on different scales (Breuer, 2002, Chap. 3). Turbulent kinetic energy is generated on the largest scales (on the order of the correlation length or boundary layer height) and is transferred down to the smallest eddies on the Kolmogorov scale $\eta = \left(v^3/\varepsilon\right)^{1/4}$, where the energy is dissipated with a rate ε into heat through viscous forcing (Stull, 1988).

The equations of motion 3.16-3.19 can only be solved numerically if all scales of turbulence are resolved. With $L/\eta \sim \mathrm{Re}^{3/4}$ orders of magnitude between the smallest and largest turbulent scale (cf. Eq. 2.6), such a direct numerical simulation (DNS) is extremely computationally expensive for large Reynolds numbers that are usual in the atmospheric boundary layer (cf. Chap. 2). It can therefore be useful to divide the full range of scales into two parts: the large scales, which describe the large scale flow, and the small scales. Splitting up each variable in Eqs. 3.16-3.19 in this way leads to governing equations for the large scale flow. Since the small and large scales are not independent, those equations are coupled to the small-scale flow via flux terms, which have to be parameterized with approximations using the large scale variables. This is known as the closure problem.

The exact position of the spectral separation depends on the scale of the atmospheric phenomena to be studied. Synoptic-scale weather forecast models use the Reynolds-Averaged Navier-Stokes-Equations (RANS), in which the scale separation is set to the atmospheric spectral gap (i.e., scales corresponding to a duration of one half to one hour, cf. Stull, 1988), to distinguish mean flow and turbulent flow. The average effect of turbulence on the mean flow is then parameterized, the structure of turbulence however cannot be investigated.

Large-eddy simulations attempt to bridge the gap between fully-resolved turbulence in DNS and unresolved turbulence in RANS-models by setting the spectral division inside the turbulent part of the spectrum. In this way, the energetically dominant turbulent scales can be resolved with computational costs considerably lower than for DNS. The smallest scales of turbulence still have to be parameterized with a subgrid-scale (SGS) model.

The separation of a variable ϕ into grid-scale part $\overline{\phi}$ and subgrid-scale part ϕ' is given by Breuer (2002):

$$\phi(\mathbf{r},t) = \overline{\phi}(\mathbf{r},t) + \phi'(\mathbf{r},t) \,, \qquad \text{[3.20a]}$$

$$\overline{\phi}(\mathbf{r},t) = \int \mathrm{d}^3 r' \, G(\mathbf{r},\mathbf{r}';\Delta) \, \phi(\mathbf{r}',t) \,, \qquad \text{[3.20b]}$$

where G is the filter kernel with characteristic filter width Δ.

The filter width is not necessarily given by the spacing of the numerical grid (cf. Breuer, 2002, Chap. 3.2). However, the present model (cf. Chap. 3.2.2) uses an implicit filtering technique by Deardorff (1970) and Schumann (1975): the spatial differentials are approximated by finite differences over the respective grid cell, which essentially means that G is constant on the grid cell and zero otherwise, i.e., G is a top-hat filter (Breuer, 2002) with Δ given by the grid spacing. This method has the advantage that it evolves naturally from the numerical method, no explicit filtering is necessary. Furthermore, $\overline{\overline{\phi}} = \overline{\phi}$, i.e. the filtered variable is not changed by further filtering, which leads to less coupling terms. LES models have been widely used to investigate boundary layer structures. An overview is given in Chap. 2.4.

3.2.2. The PALM Model

The model PALM ("A PArallelized LES Model") is a large-eddy simulation model which was developed by Raasch and Etling (1991) at the Institute for Meteorology and Climatology at Leibniz Universität Hannover, Germany, and has since been expanded and parallelized (Raasch and Schröter, 2001). Throughout the last years, several studies have proven the ability of PALM to model turbulent boundary layers. An overview of the model and published numerical studies can be found at the website of the PALM group (Raasch, 2014). The simulations used for this study were created using PALM version 3.9.

The governing equations in PALM are derived from Eqs. 3.16-3.19, using the Boussinesq-approximation (Etling, 2008), which also implies incompressibility but allows for density variations in the buoyancy term of the vertical component of the momentum equation, and a subgrid-scale parameterization according to Deardorff et al. (1980) . The governing equations for the gridscale-variables in a dry atmosphere are (Heinze, 2013):

$$\frac{\partial \overline{u}_i}{\partial t} = -\frac{\partial (\overline{u}_i \overline{u}_j)}{\partial x_j} + g\frac{\overline{\theta} - \langle \overline{\theta} \rangle}{\theta_0}\delta_{i,3} - \varepsilon_{ijk}f_j\overline{u}_k + \varepsilon_{i3k}f_3 U_{Gk} - \frac{1}{\rho_0}\frac{\partial \overline{\pi}}{\partial x_i} - \frac{\partial \tau_{ij}^r}{\partial x_j}$$

$$[3.21\text{a}]$$

$$\frac{\partial \overline{u}_i}{\partial x_i} = 0 \qquad\qquad [3.21\text{b}]$$

$$\frac{\partial \overline{\theta}}{\partial t} = -\frac{\partial (\overline{u}_j \overline{\theta})}{\partial x_j} - \frac{\partial \tau_{\theta j}}{\partial x_j} + \overline{Q}_\theta \qquad\qquad [3.21\text{c}]$$

$$\frac{\partial e}{\partial t} = -\frac{\partial (\overline{u}_j e)}{\partial x_j} - \tau_{ij}\frac{\partial \overline{u}_i}{\partial x_j} + \frac{g}{\theta_0}\tau_{\theta 3} - \frac{\partial}{\partial x_j}\left[\overline{u_j'\left(e' + \frac{p'}{\rho_0}\right)}\right] - \varepsilon \qquad [3.21\text{d}]$$

$$\overline{p} = \overline{\pi} - \frac{2}{3}\rho_0 e \qquad\qquad [3.21\text{e}]$$

where $e = \frac{1}{2}\overline{u_i'^2}$ is the subgrid-scale turbulent kinetic energy, π is the modified pressure including the contributions from the diagonal of the subgrid-scale momentum flux tensor $\tau_{ij} = \overline{u_i'u_j'}$, $\tau_{ij}^r = \tau_{ij} - \frac{2}{3}e\delta_{ij}$ is the traceless subgrid-scale momentum flux tensor, $\tau_{\theta j} = \overline{\theta'u_j'}$ is the subgrid-scale flux of potential temperature, $\mathbf{U}_G$ is the geostrophic wind vector, and ε is the diffusion rate. The ground states are $\rho_0 = 1$ kg/m^3 and θ_0 the initial temperature profile. The higher-order moments to be parameterized in a subgrid-scale model are therefore

$$\tau_{ij}^r, \quad \tau_{\theta j}, \quad \text{and} \quad \overline{u_j'\left(e' + \frac{p'}{\rho_0}\right)}. \tag{3.22}$$

PALM uses a gradient transport approach to parameterize these moments and the dissipation ε (Raasch and Etling, 1991; Heinze, 2013). Since the moments in 3.22 are of second and third order, this method is called a one-and-a-half order closure technique (Stull, 1988).

The set of equations 3.21 is complemented by equations and terms for moisture and large scale subsidence, which are neglected here because they do not contribute to the present study.

The equations 3.21 are solved numerically on a staggered Arakawa-C grid (Fig. 3.4 Arakawa and Lamb, 1977) for improved spatial resolution. Spatial differentials are approximated by finite differences (i.e., $\frac{\partial}{\partial x_i} \to \frac{\Delta}{\Delta x_i}$) for terms linear in prognostic variables, and a Wicker-Skamarock-scheme (Wicker and Skamarock, 2002) for the flux-terms. Time integration is executed with a third order Runge-Kutta scheme (cf. Baldauf, 2008). Numerically, the incompressibility condition Eq. 3.21b is preserved by a predictor-corrector method (Steinfeld, 2009): The Navier-Stokes equations 3.21a are integrated without the pressure term to obtain a preliminary solution for the u_i, which in combination with the time-integrated pressure term must yield the incompressible real

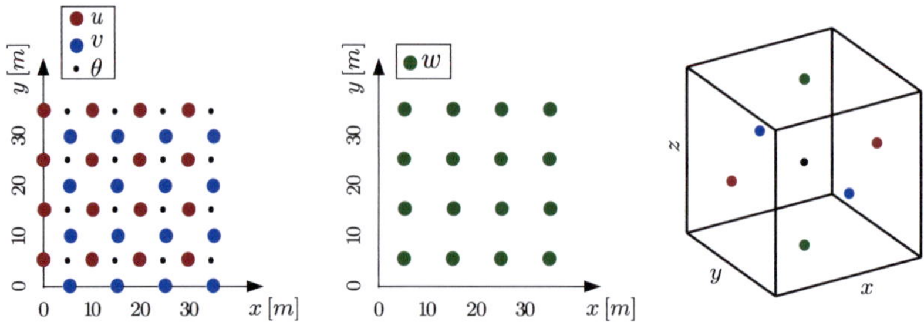

Figure 3.4.: Illustration of the staggered Arakawa-C grid used in PALM for a grid spacing $\Delta = 10$ m: The horizontal layers of u, v and scalars (left) and of w (center) are stacked alternately in vertical direction at $\Delta z/2$-intervals. The right panel shows one grid cell.

wind field. With this condition, the wind field can be determined from a Poisson-equation for the pressure, with is solved with an FFT (Fast Fourier Transform) algorithm.

To be solvable, the system of differential equations requires boundary conditions. At the start time of the simulations, all variables are pre-scribed by vertical profiles, assuming horizontal homogeneity. Laterally, cyclic boundary conditions are assumed. The boundary conditions at the top and bottom of the simulated region for the LES in this study are listed in Tab. 5.1. Between the bottom at $z = 0$ and the first layer of resolved wind field components at z_p, a Prandtl-layer is defined (Steinfeld, 2009), which allows to derive the bottom boundary values at z_p for the subgrid-scale momentum flux terms $\overline{u'w'}$ and $\overline{v'w'}$ by integrating the Businger-Dyer equations (e.g. Stull, 1988) from the roughness length z_0 to z_p. Furthermore, a constant surface heat flux is prescribed at the bottom. If the simulation is carried out over a homogeneous surface, no turbulence will develop naturally. Therefore, random disturbances with small amplitudes are superimposed on the wind fields at constant time intervals, until a steady turbulent state has developed.

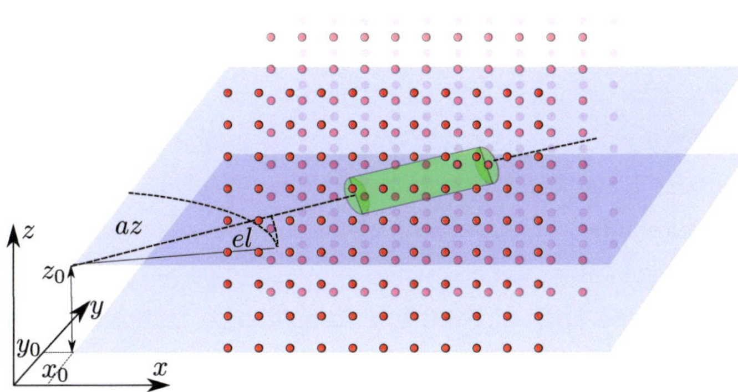

Figure 3.5.: Stawiarski et al. (2013): Geometry of the lidar simulation: The position and orientation of each range gate in the LES data grid (red) is determined from the virtual lidar position at (x_0, y_0, z_0) in the LES axes and the azimuth and elevation angles (az, el) at the time of measurement. The green cylinders with height Δp_{eff} indicates the approximate region of LES data points used for averaging with a fixed beam. ©2013 American Meteorological Society. Used with permission.

The scope of PALM reaches beyond the dry boundary layers with flat surfaces described above: the model allows to simulate moisture and clouds (Riechelmann et al., 2012), Lagrangian movement of particles (Steinfeld et al., 2008) and oceanic turbulence (Raasch and Etling, 1998). Depending on computational power, the grid spacing can reach down to 2 m (Raasch and Franke, 2011), and it is possible to include heterogeneous surfaces (Letzel et al., 2008).

In this study, PALM simulations are used for a comparison of 'real' LES boundary layer wind fields with those derived from virtual lidar measurements in the LES fields. The model set-up is described in Chap. 5.

3.2.3. Doppler Lidar Simulations based on LES

To perform virtual Doppler lidar measurements inside an LES boundary layer, a lidar simulation software package was developed.

The simulator is controlled via a text file, in which the crucial input parameters are specified: the LES data set, range gate lengths, number and positions along the lidar beam, laser pulse width, measurement frequency, lidar position and scan pattern (cf. App. B). The scan patterns are defined for a certain time interval on the LES time axes, i.e. different scan types can be performed consecutively in the same virtual lidar measurement.

The geometric information contained in the scan pattern, the range gate positions and the lidar position are combined to compute the spatial position of each range gate along the beam at each time step during the scan in the LES grid (cf. Fig. 3.5). The time axis for the velocity estimation is defined by the measurement frequency f: At time $t_k = k/f + t_{start}$, a velocity estimate is computed for each range gate using the beam position in the time interval $\Delta t(k) = [t_k - 1/(2f), t_k + 1/(2f)]$. As discussed in Chap. 3.1.1, the mathematical model for the velocity estimate in one range gate is given by a weighted average over the radial velocity along the beam, with an additional linear average over all beam positions during $\Delta t(k)$ in the case of scanning lidars.

Numerically, the velocity estimator is implemented as follows: For range gate n at distance $r_0(n)$ along the beam, the radial weighting function around the range gate center is computed (Eq. 3.6). Since the mathematical model for the weighting function is always positive, a cutoff has to be chosen: Here only the beam segment is considered on which the weighting function is larger than 20% of its maximum value at $r_0(n)$, resulting in an effective range gate length Δp_{eff}. Hereafter the area is computed which is covered by the beam segment $[r_0(n) - \Delta p_{eff}/2, r_0(n) + \Delta p_{eff}/2]$ during the time $\Delta t(k)$. This area is filled with a grid of points (cf. Fig. 3.6), to which each of the three 3D wind field components of the LES is interpolated using MATLAB built-in

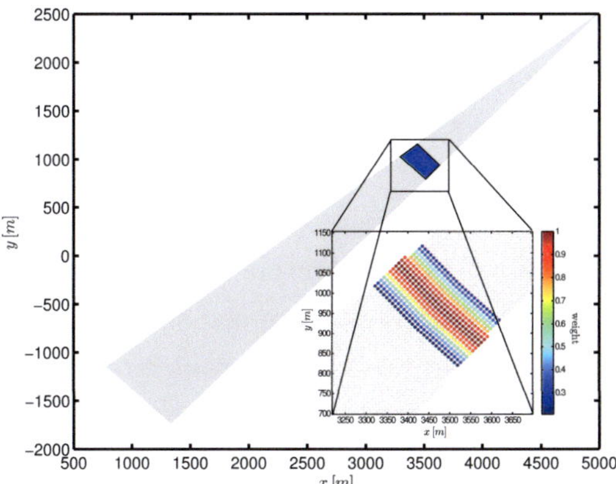

Figure 3.6.: Velocity estimation in the lidar simulator: A virtual lidar at $x = 5000$ m and $y = 2500$ m scans the horizontal plane with $\omega = 6.9°/s$. For a measurement frequency of $f = 1$ Hz, the beam covers an angle ω/f per velocity estimate (gray area). Of this section, a part of length Δp_eff is considered around each range gate center (blue area). The LES wind field components are interpolated to the local grid with relative weights according to Eq. 3.10 (inset).

cubic spline interpolation. The grid point spacing equals the LES grid constant to make full use of the LES resolution. An average value is computed for each wind field component after assigning relative weights to the interpolation points according to Eq. 3.10. The radial velocity estimate is subsequently computed as the projection of the average wind vector on the lidar beam direction. Note that averaging and projecting are independent linear operations and can be interchanged, therefore performing the wind vector component averages before the projection on the radial direction is not a source of error. This single velocity estimation is repeated for all range gates centered at $r_0(n)$, $n = 1, \ldots, \textit{range gate number}$, and all points on the time axis t_k.

The accuracy of the lidar simulator, i.e. its ability to produce realistic virtual lidar measurements in an LES boundary layer given that the mathematical models for beam averaging are correct, depends of the relation of lidar averaging length scales to the LES grid spacing L_G: When $\Delta p \gg L_G$, the average along the beam can be assumed to be realistic. Here, typical values are $\Delta p \geq 60$ m and $L_G = 10$ m. However, the real lidar beam width of approximately 10 cm means insufficient resolution in both directions perpendicular to the beam. The scanning patterns described in Chap. 5 only use fast-scanning beams, therefore only the direction normal to the scanning plane remains poorly resolved. In the following it will be assumed that the large-scale averaging in the two lidar plane directions, as well as the time-averaging involved, will smooth out all small-scale processes. Furthermore, the comparative LES data are interpolated to the scanning plane using the same technique (Chap. 5.4), thereby restoring the comparability of the results.

The lidar simulator only performs single lidar measurements. Therefore, for synchronized dual-Doppler measurements, the scan patterns of both lidars have to be planned to perform synchronized scans on the same LES data. After the two single lidar simulations are completed, both are reassembled in the dual-lidar retrieval algorithm, which works equally for both simulated and measured lidar data.

3.3. Dual-Doppler Retrieval of the Horizontal Wind Field

The data from dual-lidar scans can be used to retrieve the projection of the wind vector in the two-dimensional lidar plane.
For unsynchronized scans which take T_1 and T_2 for a full back-and-forth sector sweep, respectively, the best achievable time resolution of the retrieval is $T_0 = \max\{T_1, T_2\}$. To obtain the highest possible time

resolution, the scan should be synchronized, i.e. the beams arrive at their turning points simultaneously, thereby $T_1 = T_2$. Since no phase shift is accumulated, $T_0 = T_1/2$ in this case, which means the duration of one beam sweep is sufficient to gather dual-lidar data in the whole overlap area.

A retrieval algorithm was developed based on Newsom et al. (2008) for zero- or low-elevation scans, i.e. the retrieval plane is the horizontal plane at lidar height. A detailed description can also be found in Stawiarski et al. (2013).

The retrieval accepts real lidar measurement data or virtual lidar data created by the simulation tool (Chap. 3.2.3). Before the retrieval, real data are filtered from erroneous velocity estimates using a hard target filter (removing data points with high SNR and low absolute wind speed), an SNR filter (removing data points with low SNR) and a velocity jump filter (eliminating outliers in the time series of each range gate velocity estimate). The threshold values for the respective filters can be specified by the user.

The retrieval algorithm starts by subdividing the dual-lidar measurement time axis into time intervals of length T_0. The horizontal overlap area is then covered by a Cartesian grid with lattice constant $\Delta xy = \Delta p$.[1] Each grid point is surrounded by a circular grid cell, and all radial velocities are associated with this cell if their range gate center falls into the cell at some point during the T_0 interval. The radius of the circles is set to $R = \Delta xy/\sqrt{2}$, i.e. the smallest value to cover the whole grid (Fig. 3.7).

---

[1] Full area coverage could also be obtained by setting $\Delta xy = $ *range gate center distance*, but the lidar resolution remains limited by Δp.

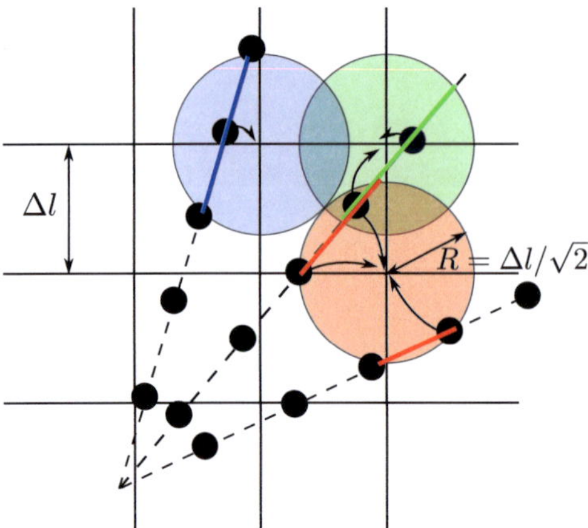

Figure 3.7: Dual lidar retrieval: A scanning lidar beam (dashed lines) with range gate centers (black bullets) moving through three exemplary grid cells. The velocity estimates are associated with the grid cells and weighted with the length of the beam chord inside the cells, indicated by the colored beam segments. Figure adapted from Stawiarski et al. (2013).

This leads to smoother results with less errors compared to a nearest-neighbor approach, since the retrieval is computed from more data.

For each grid cell around the point $\mathbf{r}_0$ and each time interval, the most probable horizontal wind vector $\mathbf{u}_H$ is then given as the minimum of the cost function (Stawiarski et al., 2013)

$$J = \sum_n g_n \left(rv_n - \mathbf{u}_H(\mathbf{r}_0) \cdot \hat{\mathbf{r}}_n \right)^2 \qquad [3.23]$$

with the radial velocities rv_n accumulated in the cell, and their associated normalized lidar beam direction vectors $\hat{\mathbf{r}}_n$. The relative contribution of the deviation of rv_n from the 'true' radial wind in this direction is additionally weighted with a factor g_n, which was not used in the original algorithm by Newsom et al. (2008). Here, g_n is set to the length of the beam segment which lies inside the cell. Thereby, the importance of data from range gate centers close to the edge of the cell is suppressed compared to those which are closer to the center of the cell. This reflects the fact that, the shorter the beam chord inside the cell, the more information from outside the cell is contained in the velocity estimate, which should not contribute to the cell result for $\mathbf{u}_H(\mathbf{r}_0)$.

The minimization of Eq. 3.23 with $\delta J = 0$ yields (Stawiarski et al., 2013):

$$\mathsf{M} \cdot \mathbf{u}_H(\mathbf{r}_0) = \mathbf{b} \,, \qquad\qquad [3.24]$$

with

$$\mathsf{M} = \sum_n g_n \hat{\mathbf{r}}_n \hat{\mathbf{r}}_n^T \qquad \text{and} \qquad\qquad [3.25a]$$

$$\mathbf{b} = \sum_n g_n r v_n \hat{\mathbf{r}}_n \,, \qquad\qquad [3.25b]$$

where $\hat{\mathbf{r}}_n$ is to be understood as a two-component column vector on the basis formed by the Cartesian horizontal axes:

$$\hat{\mathbf{r}}_n = \begin{pmatrix} \sin(az_n)\cos(el_n) \\ \cos(az_n)\cos(el_n) \end{pmatrix} . \qquad\qquad [3.26]$$

Eq. 3.24 is solved for each grid cell in each time interval, thereby the horizontal wind field is obtained for the whole overlap region and measurement time. Note that the matrix M is not invertible if all $\hat{\mathbf{r}}_n$ are equal save for a scalar factor, i.e. the lidar beams must not be collinear in the grid cells.

Newsom et al. (2008) used a similar retrieval algorithm for the investigation of surface layer coherent structures in dual-Doppler lidar measurements during the Joint Urban 2003 (JU2003) field campaign. They achieved a time resolution of $T_0 = 30$ s and a spatial resolution of $\Delta xy = 100$ m with the unsynchronized lidars. Hill et al. (2010) adapted the algorithm for the retrieval in vertical planes during the Terrain-Induced Rotor Experiment (T-REX) with $T_0 = 40\text{-}50$ s and $\Delta xy = 130$ m.
Iwai et al. (2008) attempted a three-dimensional retrieval of the wind field, which required PPI-scans at different elevation angles for each retrieval interval. Therefore, they only realized a time resolution of $T_0 = 12$ min and $\Delta xy = 100$ m.

During the HOPE-experiment (cf. Chap. 8 for an overview), the KIT dual-Doppler lidar system achieved a time resolution of $T_0 = 12$ s and a spatial resolution of $\Delta xy = 60$ m for horizontal scans.

In realistic lidar set-ups, it is hardly ever possible to perform zero-elevation coplanar scans: hard targets such as houses, trees or even transmission towers can block the beam path. The former can often be avoided using a small elevation, which leads to slightly tilted lidar planes. The consequences, as well as error contributions from angles, random noise, time undersampling etc. are discussed in Chap. 4, where an optimization scheme is developed for error reduction.

4. Errors in Dual-Doppler Lidar Measurements

The following error analysis of dual-Doppler lidar measurements, as well as the associated appendix chapter (App. A), are an excerpt from the publication
Stawiarski, Träumner, Knigge, and Calhoun, 2013: *Scopes and Challenges of Dual-Doppler Lidar Wind Measurements - An Error Analysis.* J. Atmos. Ocean. Tech., **30(9)**, 2044-2062. ©*2013 American Meteorological Society. Used with permission.*

4.1. Error Sources in Dual-Doppler Lidar Measurements

The usage of dual-Doppler system accounts for several errors. In this section, we discuss the errors listed in Tab. 4.1 and their relative influence for the different scan types of Chap. 3.1.3. App. A contains the detailed error propagation of single lidar errors to dual-Doppler results. As a convention, we write $\langle \xi \rangle_n$ for the nth moment of any variable ξ, i.e. $\langle \xi \rangle_1$ is the expectation value and $\langle \xi \rangle_2$ the variance.

4.1.1. Single Lidar Random Errors

The measured radial velocity of a Doppler lidar is typically described as follows (Frehlich, 2001; Davies et al., 2005):

$$rv_i^M(R_0,t) = rv_i(R_0,t) + \varepsilon_i(R_0,t) + b_i^{rv}(R_0,t) \qquad [4.1]$$

Error	Symbol	Source
Single Lidar Errors		
Single lidar uncorrelated noise (random error)	$\sigma_i^{rv,rnd}$	Random measurement inaccuracy due to speckle effect, detector noise (Frehlich, 2001).
Single lidar bias (systematic error)	$bias_i^{rv,est}$	Measurement bias due to frequency drift of the laser, nonlinear amplifiers, digitization errors, non-ideal noise statistics (Frehlich et al., 1994).
Direction errors	σ_i^{az}, σ_i^{el}, $bias_i^{az}$, $bias_i^{el}$	The azimuth and elevation angles are slightly imprecise, due to an imperfect adjustment of the lidar systems and/or the moving of the scanner.
Derived Single Lidar Errors		
In-plane error	$\sigma_i^{rv,ip}$, $bias_i^{rv,ip}$	Direction errors, projected on lidar plane, lead to line-of-sight velocity estimation errors.
Out-of-Plane error	$\sigma_i^{rv,oop}$, $bias_i^{rv,oop}$	Direction errors perpendicular to lidar plane lead to errors in rv that scale with perpendicular wind speed.
Dual Lidar Errors		
Single Lidar Propagated Error	σ_{DD}^{single}, $bias_{DD}^{single}$	Propagation of single lidar errors to dual-Doppler result.
Time Averaging Error	σ_{DD}^{T}	Data from both lidars is not synchronous or does not cover full/same retrieval time.
Volume Error	σ_{DD}^{V}	Both lidar beams cover different/large volumes of air, due to scan/beam separation/beam direction.

Table 4.1.: Overview of the occurring errors in dual-Doppler lidar measurements

for a range gate centered around R_0. $b_i^{rv}(R_0,t)$ is a systematic error with the assumptions $\langle |b_i^{rv}|\rangle_1 = |bias_i^{rv,est}|$, and $\langle bias_i^{rv}\rangle_2 = 0$ and $\varepsilon_i(R_0,t)$ is a random error with $\langle \varepsilon_i\rangle_1 = 0$ and $\langle \varepsilon_i\rangle_2 = \sigma_i^{rv,rnd}$ (see Tab. 4.1 for sources).

For the error analysis of the single lidar systems, measurements against the 200 m tower (Barthlott et al., 2003) at KIT, Campus North, were performed. Both data sets were obtained when the lidar systems first became operational, i.e. for 'WindTracer 1' (2 μm) from January 19 to 24, 2005 and for 'WindTracer 2' (1.6 μm, cf. Tab. 3.1) from January 26 to 31, 2011. The laser beams were arranged in a way that the centers of the $10th$ range gates were located near sonic anemometers located in 100 m and 200 m height at the tower.

The local morphology shows low-rise buildings in the first 1.5 km along the laser beam and a forest area behind. During the test measurements in 2005 a 50 pulse average was applied (resulting in a measurement rate of 10 Hz), during 2011 a 75 pulse average was used from 24 to 27 January (10 Hz measurement rate) and a 750 pulse average afterwards (1 Hz measurement rate).

To evaluate the systematic error the line-of-sight velocity measured by the lidar was compared with the line-of-sight projection of the wind vector measured by the sonic anemometer. However, the bias between the two measured wind velocities depends strongly on the wind direction, i.e. there seem to be strong effects by the tower which render the used method inapplicable. From the unperturbed areas we derive $bias_i^{rv,est} \leq$ 0.2 m/s, $i = (1,2)$.

To estimate the uncorrelated noise, the technique based on the difference between lag zero and and lag one of the autocorrelation function was used (Lenschow et al., 2000). This procedure leads to a slight overestimation of the uncorrelated noise but is more robust than fitting techniques or the use of the spectra. The autocorrelation function was calculated for 30 min time intervals. Fig. 4.1 shows the results for both lidar systems. The used SNR has a 6 MHz bandwidth. A strong increase in

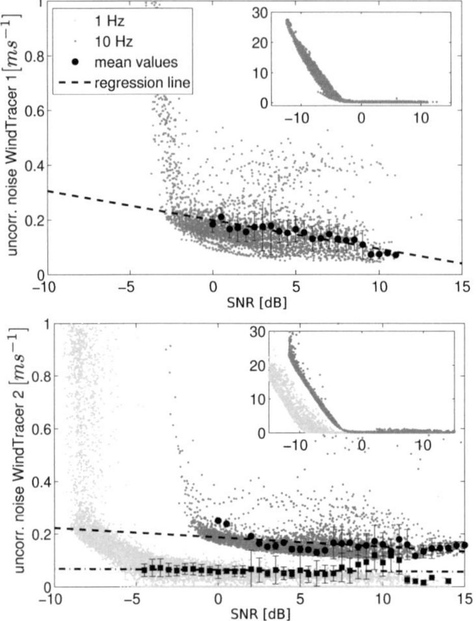

Figure 4.1: Uncorrelated noise of the two KIT lidar systems as a function of signal-to-noise ratio. At the top, the results for 'WindTracer 1' and at the bottom, the results for 'WindTracer 2' are shown. Light and dark gray dots denote 10 Hz and 1 Hz measurement frequency, respectively. The black dots and error bars are SNR bin means and standard deviations. The dashed lines are linear fits to the bin means.

the uncorrelated noise is visible at around -2 dB for the 'WindTracer 1' and 'WindTracer 2' when averaging 50 and 75 pulses, respectively. If 750 pulses were averaged the noise increases at about -8 dB. For SNR higher than the given thresholds, the uncorrelated noise $\sigma_i^{rv,rnd}$ is below 0.2 m/s.

4.1.2. Single Lidar Direction Error

Systematic direction errors occur if the lidar systems are not set-up properly. We mostly estimate that the azimuth and elevation direction are precise up to a bias of about 0.2°, denoted here as $bias_i^{az/el}$. The scanner is aligned by detecting hard-target backscatter signals from far-away objects, and estimates of the biases can be derived from known accuracies of lidar and object positions and the statistical accuracy of the scanner. Measurements using repeated scanner movements to this alignment position show that the scanner does not accumulate further bias when scanning, and that the statistical errors $\sigma_i^{az/el}$ in both angular directions are smaller than $0.1°$.

Depending on the measurement, the movement of the scanner during the velocity estimation can be seen as either a desired feature, or an additional source for statistical errors. For a sampling frequency of 10 Hz and a scan velocity of $5°$ per second, the angle is never better located than $0.5°$, which corresponds to a spatial interval of about 17 m in a distance of 2 km. Some scan patterns make it necessary to tilt the lidar beam away from the desired direction, e.g. to avoid obstacles. Both tilted and inaccurate lidar beams lead to faulty velocity estimations, since even if the LOS velocity estimator were free of errors, an incorrect wind field component is sampled. To estimate the magnitude of these errors, we define an *evaluation plane* as the plane in which we want to retrieve the two-dimensional wind field, regardless of the actual lidar plane. The radial velocities in the evaluation plane are then perturbed by statistical errors and biases due to instrument errors and beam tilt in or away from this evaluation plane.

For convenience, the angular errors are split into two parts, the in-plane and out-of-plane error. This splitting is advisable, since the in-plane errors scale with the in-plane wind velocity, which can be retrieved from measurement data, whereas the out-of-plane errors scale with the plane-normal part of the wind vector, and therefore their estimation requires additional measurements with other equipment (cf. App. A).

Using this, the errors in Eq. 4.1 become

$$(\sigma_i^{rv})^2 = \left(\sigma_i^{rv,rnd}\right)^2 + \left(\sigma_i^{rv,ip}\right)^2 + \left(\sigma_i^{rv,oop}\right)^2 + \text{cov}(ip, oop) \qquad [4.2]$$

and

$$|bias_i^{rv}| = |bias_i^{rv,est}| + |bias_i^{rv,ip}| + |bias_i^{rv,oop}| \ . \qquad [4.3]$$

In App. A, formulae for the statistical errors and biases are derived. They are functions of the wind speed (in-plane and out-of-plane), wind direction, plane orientation and angle between lidar beams. The symbols for these parameters are introduced in Chap. 3.1.3 and summarized in Fig. 3.3.

To estimate the magnitude of the in-plane error, we consider as an example the configuration of a ground-parallel plane, $\hat{\mathbf{k}} \times \hat{\mathbf{n}}_n = \mathbf{0}$, with $\gamma_z \in \{0°, 180°\}$, and $el_1 = el_2 = 0$. In App. A, the upper bounds of the variance and bias were derived:

$$\left(\sigma_i^{rv,ip}\right)^2 = u_H^2 \sin^2\left(\alpha_j \mp \frac{\Delta\chi}{2} - \gamma_{u_H}\right)(\sigma_i^{az})^2$$

$$\leq u_H^2 \left(\sigma_i^{az}\right)^2 \qquad [4.4a]$$

$$\left|bias_i^{rv,ip}\right| = |u_H| \left|\sin\left(\alpha_j \mp \frac{\Delta\chi}{2} - \gamma_{u_H}\right)\right| |bias_i^{az}|$$

$$\leq |u_H| |bias_i^{az}| \ . \qquad [4.4b]$$

For a wind speed of $u_H = 5$ m/s and the direction error estimations above, the statistical error therefore has an upper bound of 0.01 m/s and the bias has an upper bound of 0.02 m/s. Both are one order of magnitude smaller than the statistical error in rv due to instrument noise and can thus be neglected. However, the in-plane errors becomes relevant if higher wind speeds, in-plane angular biases or statistical errors occur.

The out-of-plane error is the counterpart to the in-plane error and arises if the desired lidar plane is not the actual retrieval plane given by the span of the two lidar beams. This occurs if one of the lidar beams is tilted slightly away from the plane, which leads to an undesired contribution of the perpendicular wind component. The results are given in the App. A. Analogous to the in-plane case, we give an example for a a horizontal planar scan:

$$\left(\sigma_i^{rv,oop}\right)^2 = w^2 \left(\sigma_i^{el}\right)^2 \qquad [4.5a]$$

$$\left|bias_i^{rv,oop}\right| = |w| \left|bias_i^{el}\right| \; . \qquad [4.5b]$$

Since $\langle w \rangle$ is approximately zero (depending on the measurement time scale), a horizontal planar measurement is only influenced by an additional statistical variance which scales with w^2. According to Kaimal and Finnigan (1994), the vertical wind velocity standard deviation in the surface layer follows Monin-Obukhov similarity with

$$\phi_w = \frac{\sqrt{w'^2}}{u_*} , \qquad [4.6a]$$

$$\phi_w = \begin{cases} 1.25(1+3|z/L|)^{1/3}, & -2 \leq z/L \leq 0 \\ 1.25(1+0.2|z/L|) , & 0 \leq z/L \leq 1 \end{cases} , \qquad [4.6b]$$

with u_* the friction velocity and L the Obukhov length. The horizontal planar scans in the JU2003 study (Newsom et al., 2008) used elevations of $0.5°$ and $1.2°$, and the stability conditions lead to out-of-plane error contributions of $\sigma^{rv,oop} = 0.01 \; m/s$ to $0.05 \; m/s$, which is small compared to the random instrument error. The out-of-plane error should be computed nevertheless for every scan to assure its negligibility, since it depends on atmospheric conditions. The covariance between the in-plane and the out-of-plane error, $\mathrm{cov}(ip,oop)$, is zero for the given example (cf. App. A). This contribution to the error becomes only relevant for evaluation planes which lie neither parallel nor perpendicular to the ground.

4.1.3. Dual Lidar Propagation Errors

In the most general case, we are interested in the wind field component $u_j = \hat{\mathbf{e}}_j \cdot \mathbf{u}$ in a certain direction $\hat{\mathbf{e}}_j$ in the evaluation plane. The errors of the individual lidar instruments will sum according to Eqs. 4.2 and 4.3. The total errors propagate to the retrieved u_j (see Eq. 3.15), using Gaussian error propagation (Hill et al., 2010):

$$\left(\sigma_{DD}^{\text{single}}(u_j)\right)^2 = \left(\frac{\partial u_j}{\partial rv_1}\sigma_1^{rv}\right)^2 + \left(\frac{\partial u_j}{\partial rv_2}\sigma_2^{rv}\right)^2$$

$$= \frac{\sin^2(\alpha_j + \frac{\Delta\chi}{2})}{\sin^2(\Delta\chi)}(\sigma_1^{rv})^2 + \frac{\sin^2(\alpha_j - \frac{\Delta\chi}{2})}{\sin^2(\Delta\chi)}(\sigma_2^{rv})^2 \ . \qquad [4.7]$$

Using the simplification $\sigma_1^{rv} = \sigma_2^{rv} = \sigma^{rv}$, which holds for identically constructed Doppler lidar systems, this simplifies to

$$\left(\sigma_{DD}^{\text{single}}(u_j)\right)^2 = \frac{\sin^2(\alpha_j + \frac{\Delta\chi}{2}) + \sin^2(\alpha_j - \frac{\Delta\chi}{2})}{\sin^2(\Delta\chi)}(\sigma^{rv})^2 \ . \qquad [4.8]$$

The pre-factor in Eq. 4.8 is illustrated in Fig. 4.2. While it is obvious that collinear beams can only resolve the wind direction in which they both point, it should be noted that a twenty degree angle $\Delta\chi$ between the lidar beams will still lead to four times the single radial velocity error for a wind field direction orthogonal to the lidar beams ($\alpha_j = 90°$). Only very few angular combinations can lead to a decrease in error (white regions in Fig. 4.2). The best achievable result for one wind field component is an error halving, but it is accompanied by a high error increase in the orthogonal component. An optimal result for two orthogonal wind field components is given for a ninety-degree angle between lidar beams, in this case the factor for both components is one.

For two perpendicular velocity components, we find that

$$\left(\sigma_{DD}^{\text{single}}(u)\right)^2 + \left(\sigma_{DD}^{\text{single}}(v)\right)^2 = \frac{(\sigma_1^{rv})^2 + (\sigma_2^{rv})^2}{\sin^2(\Delta\chi)} \ , \qquad [4.9]$$

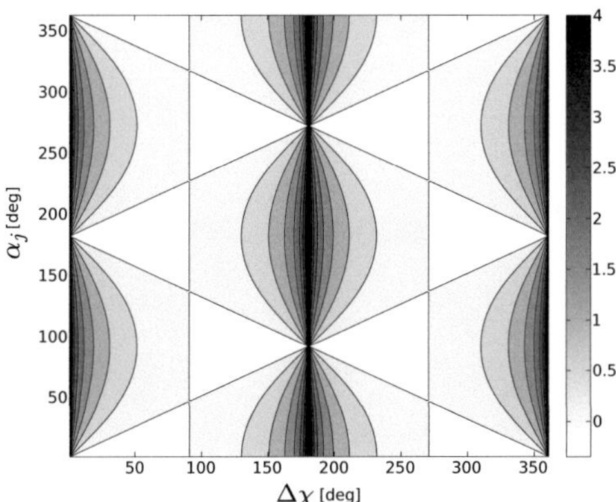

Figure 4.2: Logarithm of error-magnifying pre-factor of Eq. 4.8, i.e. $\log((\sin^2(\alpha_j + \frac{\Delta\chi}{2}) + \sin^2(\alpha_j - \frac{\Delta\chi}{2}))^{1/2}/|\sin(\Delta\chi)|)$, as a function of the angle $\Delta\chi$ between lidar beams and the angle α_j between the desired retrieved wind component direction $\mathbf{e}_i$ and the mean lidar beam direction. The pre-factor diverges in the black areas.

which also holds for $\sigma_1^{rv} \neq \sigma_2^{rv}$. This is equivalent to the results for the dual-Doppler radar application in Davies-Jones (1979). The crucial pre-factor $1/|\sin(\Delta\chi)|$ is mapped in Fig. 4.3. We find that this pre-factor can lead to extremely high errors on the retrieval results. Fig. 4.3 can help to plan scan patterns where the expected errors on the retrieval results are reasonably small. App. A shows that the propagated bias also scales with this pre-factor.

It is important to note that planar scan patterns, which use $N > 1$ velocity estimates per lidar for one grid cell retrieval, exhibit a reduced statistical variance by the factor $N/2$ compared to the intersecting beam case. This is due to the higher statistical certainty (Bronstein et al., 2001).

4.1.4. Dual Lidar Time Averaging Error

Time averaging errors are defined as errors that occur if both lidars do not provide data at the same time interval, or not for the full time interval of one retrieval.

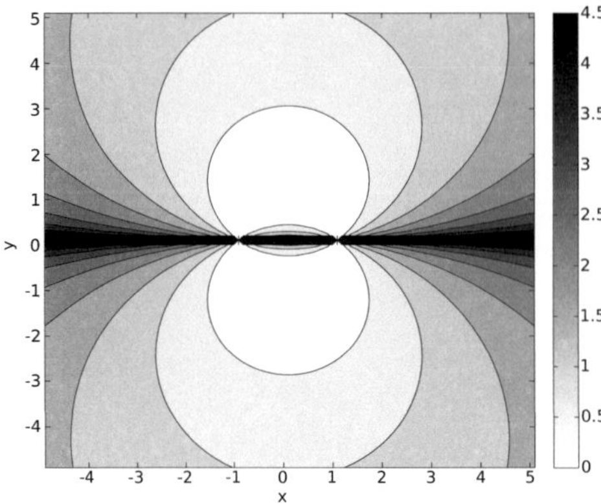

Figure 4.3: Logarithm of the error-magnifying factor, $\log(1/|\sin(\Delta\chi)|)$, for two lidars positioned at $(x,y) = (1,0)$ and $(-1,0)$. The pre-factor diverges in the black areas.

For measurements in dual-Doppler mode with high time resolution (cf. Chap. 3.1.3) it is recommended that the time standard does not differ between the two instruments, i.e. the system clocks have to be synchronized. Time shifts become a problem when regarding turbulence, and the correlation terms between the two measured radial wind velocities become apparent. The importance of synchronization errors in the intersecting beam case decreases if the time interval considered becomes longer. In the planar scan method (Chap. 3.1.3), where data is aggregated in one grid cell during time T_0, synchronization is much less crucial. However, in this method appears a different time averaging error: the temporal undersampling error.

During the time interval T_0 (cf. Chap. 3.3), each lidar beam passes a grid cell only once for synchronized systems, and the velocity estimates recorded are supposed to represent the complete time interval. The temporal undersampling error arises because of the deviation of the T_0-mean of the measured radial velocities from the desired T_0-mean. Since we do not have continuous wind speed measurements in the grid cells, we measure the fluctuation by the change in rv in each range gate

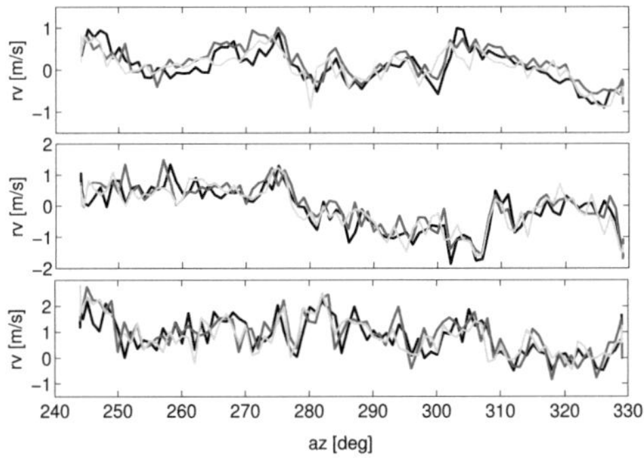

Figure 4.4.: rv time series during a sector PPI scan performed with 'WindTracer 2' on 24 Sep, 2011 in Hatzenbühl, Germany. The depicted data shows rv as a function of the azimuth angle az during three consecutive scans of the angle sector (1st scan: black, 2nd scan: dark gray, 3rd scan: light gray). The scan was performed with $T_0 = 10.5$ s, $\Delta p = 30$ m and a constant elevation of $2°$.

and each angle from one beam sweep to the next.

Fig. 4.4 shows the time series of three range gates for three consecutive sweeps as a function of the azimuth angle (elevation was kept at $2°$). The data was recorded with 'WindTracer 2' on Sep 24, 2011. It is obvious that the contribution of small scale processes leads to variations of the radial wind velocity and thus considerable time undersampling errors in the grid cells. Indeed, the average over all absolute rv-changes is 0.35 m/s for these three sweeps, which makes it the highest error contribution to planar scans.

In Chap. 4.2, we will investigate the dependence of this error on the mean wind and the range gate length, and develop a method to decrease it.

4.1.5. Dual Lidar Volume Error

Volume errors arise because radial velocity estimates used for retrievals are always averages over a certain volume of air, which is not the same volume for both lidars or not an average over the desired volume. In situations with strong tilt or strong shear this may become a problem for the dual-Doppler application, because the velocity field changes quickly and the volumes of the two lidars see may contain different turbulent structures.

In many dual-Doppler scan scenarios there are displacements between the range gate centers whose radial velocities are used for retrieval. These may arise from non-perfect synchronization but also to avoid hard targets, as discussed in Chap. 3.3. Often the wind field is considered horizontally homogeneous (Stull, 1988), i.e. we can neglect effects due to horizontal shifts. To estimate the effects in the vertical, the current wind profile may give a reference point. In general, we assume the displacement effect to be negligible if the beam displacement is much smaller than the range gate length, so that turbulence structures of the displacement scale can be assumed to average out. The error cannot be quantified without additional measurements.

For intersecting beams, the inherent spatial averaging property of the lidar becomes the source of a volume error: While the desired retrieval result is the wind field at the beam intersection point, the range gates stretch much further. This deviation from a point measurement can be estimated by the velocity fluctuations inside the range gate. Frehlich (1997) finds that, for homogeneous and isotropic flow and the Kolmogorov model structure function, these fluctuations are given by

$$(\sigma_i^V)^2 = \frac{9}{40}C_v\varepsilon^{2/3}(\Delta p)^{2/3} \qquad [4.10]$$

where $C_v \approx 2$ is the Kolmogorov constant and ε the dissipation rate, leading to $\sigma^V \approx 0.3$ m/s for $\varepsilon = 10^{-3}$ m^2/s^3 and $\Delta p = 100$ m. The influence of this error is most important for high time resolutions, whereas it can become negligible if the time scale is long enough to average out all scales up to the range gate length.

Even though the spatial averaging does not lead to a bias in the velocity estimation, the suppression of the small scales in the turbulence spectra means that momentum fluxes and variances computed from lidar measurements will usually underestimate the real value (Mann et al., 2010).

In contrast, the spatial average contained in the velocity estimates is a desired feature of planar scans. Here, the intended result is, for each grid point, to measure the wind speed in the evaluation plane, averaged over the grid cell with radius $R = \Delta p/\sqrt{2}$ and averaged over the time interval T_0. Volume errors arise from uneven line averaging weights inside and non-zero weights outside of the grid cells.

In the retrieval method in Chap. 3.3, data from both lidars is aggregated and evaluated at the same time. Note that this method is similar to first producing one average radial velocity for each lidar with the help of weights g_n, and subsequently minimizing Eq. 3.23 with only these two mean radial velocity entries. We will take this point of view here to simplify the error analysis.

The grid cell mean radial velocity from one lidar is the result of spatial averaging of the real radial velocity field in and around the grid cell centered at $\mathbf{x}_c$ with a weighting function $W(\mathbf{x}; \mathbf{x}_c)$.

If a range gate center transverses the grid cell during scanning, each velocity estimate is a product of the radial averaging process given by Eq. 3.5. This implies that velocities outside the grid cell are taken into account. This effect intensifies if the overlap the lidar beam has with the grid cell decreases. Therefore, individual weights g_n are chosen as the length of the lidar beam line segment that lies within the circle of radius R: $g_n = 2(R^2 - s_{\mathbf{x};\mathbf{x}_c}^2)^{\frac{1}{2}}$ (cf. Fig. 4.5).

For a grid cell with center $\mathbf{x}_c$, which is passed through by a range gate center in an approximately straight line at distance D from the center, the overall weight function at point $\mathbf{x}$ in the plane is then given by

$$W(\mathbf{x};\mathbf{x}_c) = \begin{cases} W_{\Delta p}(r_{\mathbf{x};\mathbf{x}_c}) \cdot g_n(s_{\mathbf{x};\mathbf{x}_c}) & , \quad \frac{s_{\mathbf{x};\mathbf{x}_c}}{\sqrt{R^2 - D^2}} \leq 1 \\ 0 & , \quad \frac{s_{\mathbf{x};\mathbf{x}_c}}{\sqrt{R^2 - D^2}} > 1 \end{cases} \qquad [4.11]$$

with the single lidar weighting function $W_{\Delta p}$ as given in Eq. 3.6.

W is furthermore a function of R, Δp and D. $r_{\mathbf{x};\mathbf{x}_c}$ is defined as the distance of $\mathbf{x}$ from the range gate center in lidar beam direction $\hat{\mathbf{r}}_i$, whereas $s_{\mathbf{x};\mathbf{x}_c}$ is the corresponding distance in direction perpendicular to the lidar beam as shown in Fig. 4.5:

$$r_{\mathbf{x};\mathbf{x}_c} = (\mathbf{x} - (\mathbf{x}_c + D\hat{\mathbf{r}}_i)) \cdot \hat{\mathbf{r}}_i \qquad [4.12a]$$

$$s_{\mathbf{x};\mathbf{x}_c} = |(\mathbf{x} - \mathbf{x}_c) \times \hat{\mathbf{r}}_i| \ . \qquad [4.12b]$$

For comparison with the ideal weighting function, W must be normalized:

$$W_n(\mathbf{x};\mathbf{x}_c) = W(\mathbf{x};\mathbf{x}_c) / \int_{\text{cell}} d^2 x\, W(\mathbf{x};\mathbf{x}_c) \ . \qquad [4.13]$$

The ideal weighting function for the velocity estimate of a grid cell is $W_0(\mathbf{x};\mathbf{x}_c)$,

$$W_0(\mathbf{x};\mathbf{x}_c) = \begin{cases} 1/(\pi R^2) & , \quad |\mathbf{x} - \mathbf{x}_c| \leq R \\ 0 & , \quad |\mathbf{x} - \mathbf{x}_c| > R \end{cases} \qquad [4.14]$$

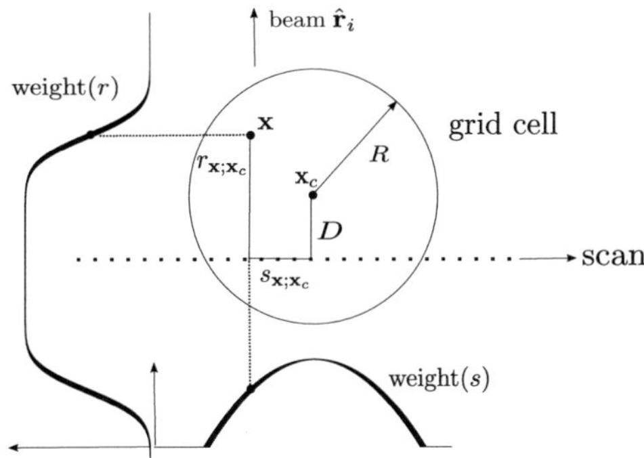

Figure 4.5: Modeled weighting of data at grid cell point $\mathbf{x}$. The lidar range gate center moves through the cell along the dotted line, at a displacement D from the grid cell center. The weights have a radial part depending on $r_{\mathbf{x},\mathbf{x}_c}$ and a cross-radial part depending on $s_{\mathbf{x},\mathbf{x}_c}$, where $r_{\mathbf{x},\mathbf{x}_c}$ is the displacement from the center in beam direction and $s_{\mathbf{x},\mathbf{x}_c}$ is the displacement in cross-beam direction (see text).

which is a solely function of R. The influence of the spatial averaging error can therefore be estimated as

$$\left(\sigma_{DD}^{V,sa}\right)^2 = \left(\int\limits_{\text{cell}} d^2x \left(W_n(\mathbf{x};\mathbf{x}_c) - W_0(\mathbf{x};\mathbf{x}_c)\right) rv(\mathbf{x})\right)^2 \qquad [4.15]$$

where $rv(\mathbf{x})$ is the real radial wind velocity.

To estimate this error, we produce random test fields with normal distribution centered around means of -20 m/s to 20 m/s, and with standard deviations of 0.5 m/s to 5.5 m/s. Using these, W and W_0 were computed for range gate lengths between 30 m and 140 m and randomly distributed distances $D = 0.05\,R$ to $D = 0.95\,R$, with $R = \Delta p/\sqrt{2}$ and $\sigma_\tau = 300$ ns. Fig. 4.6 shows the maximum value of $\sigma_{DD}^{V,sa}$ for all test wind fields as a function of Δp. It is obvious that the volume error increases, i.e. the real weighting function exhibits stronger deviations from the desired averaging, for smaller range gates. This is a result of the beam weighting function $W_{\Delta p}$ approaching the Gaussian pulse envelope for range gate lengths much smaller than the pulse width $\sigma_\tau c$. This leads to non-negligible contributions from outside the cell and to an uneven weighting of data inside the cell.

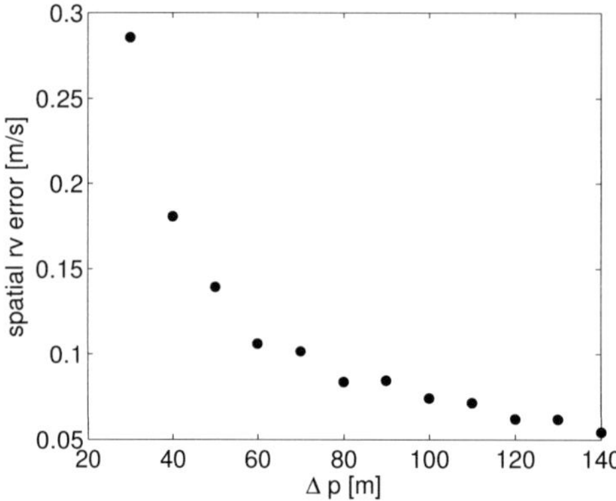

Figure 4.6: The mean error in radial velocity estimation in one grid cell due to non-optimal weighting.

The spatial error contribution is still on the order of the random error in the realms of realistic range gate lengths. For shorter range gate lengths it is advisable to choose the grid cell radius R larger to achieve a smaller spatial averaging error.

4.1.6. Summary

The relevant error processes are threefold: (i) the single lidar errors, i.e. the lidar random error and the in-plane and out-of-plane error, which influence every line-of-sight velocity estimate, and (ii) the spatial and (iii) temporal averaging errors, which arise from inaccurate spatial and temporal averaging depending on time- and length scales of the scan.

During planar scans, the single lidar errors are often negligible since the number of samples N which are used for one retrieval leads to a decrease in the standard deviations by the factor $1/\sqrt{N}$. The dominant errors contributions here come from the time and spatial averaging errors, although the latter only becomes important for very small Δp.

On the other hand, the intersecting beam case is dominated by propagated single lidar errors and spatial averaging errors. The latter arise because the averaging volume for each velocity estimate is much larger than beam intersection volume. Time averaging errors can also become important if the lidars are not properly synchronized.

All errors have to be propagated to the retrieval result in the lidar plane using Eq. 4.8 with the appropriate value for $\Delta\chi$ for each grid cell or intersection point. Figs. 4.2 and 4.3 can be used to estimate the error magnification for certain retrieved wind field components in the lidar overlap area. Scan patterns should be planned accordingly.

4.2. Optimization of Horizontal Scan Patterns

Chap. 4.1.4 showed that the dominant error process in planar scans is the undersampling of the radial velocity inside each grid cell during the time interval T_0, i.e., the few velocity estimates in each grid cell cannot reliably reproduce the mean velocity. Indeed, for Gaussian random variable with mean μ and variance σ^2, the mean computed from N samples will be Gaussian as well, with mean μ and a variance of σ^2/N (Bronstein et al., 2001). It is not possible to significantly increase the number of rv samples in the grid cells, since the time during which the velocity is sampled in one grid cell always remains only a small fraction of T_0. Consequently, we have to decrease the variance of the radial velocities in the T_0-interval. To do so, we assume a cartesian grid covering the evaluation plane with a lattice constant of $\Delta l = \Delta p$ for highest possible spatial resolution, and accordingly a radius of influence of $R = \Delta p/\sqrt{2}$ (cf. Chap. 3.3). By Taylor's hypothesis the turbulence elements can be regarded to be frozen in space and advected with the mean wind. The characteristic length scale of the measurement is the distance the wind field is transported in T_0: $\lambda = \bar{u} \cdot T_0$, where $\bar{u}$ is the mean wind velocity. If $\lambda \gg \Delta p$, major turbulence elements will not be averaged out by the lidar averaging process, which only covers scales up to Δp. Therefore the wind speed variance of those elements will be visible in the rv time series. On the other hand, if $\lambda \ll \Delta p$ the variance will be small, because all turbulence elements of the scale λ are averaged out by the radial velocity measurement process. We can therefore minimize the time sampling error by minimizing T_0 (and consequently the length scale λ) with respect to Δp.

As a cutoff, we demand

First Optimization Condition: Small Sampling Error

$$T_0 \cdot \bar{u} \leq \Delta p \, . \qquad\qquad [4.16]$$

A very high angular scan velocity could therefore theoretically solve this time sampling problem, if it were not connected with decreasing data density: If we attempt to cover an area with maximum distance d from the lidar, the distance between two consecutive velocity estimates in the outermost range gate centers must still be smaller or equal to $\Delta l = \Delta p$, to ensure at least one velocity estimate from each lidar in each grid cell. For a fixed measurement frequency f and the full angle β covered in T_0 (i.e., twice the angle sector for a back-and-forth sweep scan), this can be summed up in the

Second Optimization Condition: Sufficient Data Density

$$T_0 \cdot \Delta p \geq C_s \qquad \text{with } C_s = \frac{2\pi\beta d}{f 360°} \, . \qquad [4.17]$$

A dual-Doppler lidar scan pattern is optimized if T_0 and Δp are chosen from all parameter pairs that fulfill both conditions in a way that minimizes Δp. From the above equations we deduce the optimized values, keeping in mind that Δp has an effective lower bound of $\Delta p_{min} = \sqrt{\log 2}\, \sigma_\tau c$, which is the full width at half maximum for the spatial pulse envelope I_n (Eq. 3.8):

Ideal Scan Parameters

$$\text{If } \sqrt{\bar{u}C_s} \geq \Delta p_{\min} : \quad \Delta p_{\text{opt}} = \sqrt{\bar{u}C_s}$$
$$T_{0_{\text{opt}}} = \sqrt{C_s/\bar{u}}$$
$$\text{If } \sqrt{\bar{u}C_s} < \Delta p_{\min} : \quad \Delta p_{\text{opt}} = \Delta p_{\min}$$
$$T_{0_{\text{opt}}} = C_s/\Delta p_{\min}$$

$$[4.18]$$

In the following, we will demonstrate the effect of optimization on the time sampling error.

4.2.1. Lidar Data Results

Lidar measurements were performed by the KIT 'WindTracer 2' with a wavelength of $\lambda = 1.6\ \mu m$ (cf. Chap. 3.1.2). We used 75 pulses to average for each velocity estimate, which means a measurement frequency of 10 Hz. Data was recorded on 8 days in fall 2011 and spring 2012, on which the lidar was positioned on farmland in Hatzenbühl, Germany. The scanning area mainly covered fields with different crops, interspersed with few bushes and trees, therefore an elevation of two or three degrees had to be used. The scanned area spans an azimuthal range of around $100°$ around the west/northwest direction. Additional wind speed data was retrieved as 10 min averages from a 20 m-tower anemometer, located approximately 10 m next to the lidar. This measured wind speed can be seen as representative for the mean wind speed in the scanning area over this mostly homogeneous terrain.

We define T_0 as the full back-and-forth scan time. Since T_0 and Δp were kept relatively constant for all measurements while the mean wind $\bar{u}$ varied, we introduce the relative optimization parameters:

$$f_p = \Delta p/\Delta p_{\text{opt}} , \qquad [4.19a]$$
$$f_T = T_0/T_{0_{\text{opt}}} . \qquad [4.19b]$$

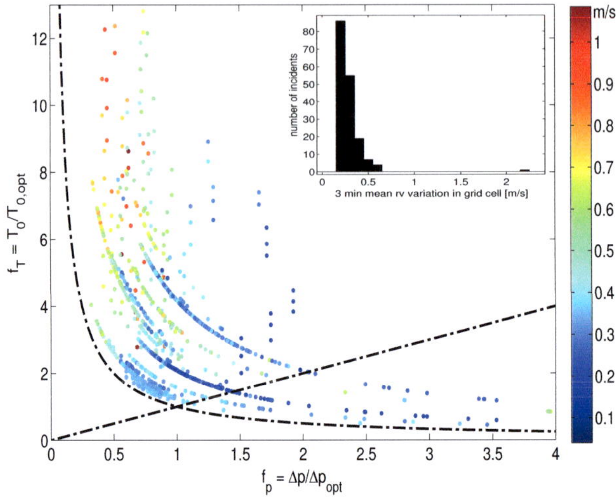

Figure 4.7.: Time undersampling error: Mean radial velocity changes in 'WindTracer 1' lidar data from one sweep to the next, averaged over all grid cells of the Δp-spaced grid and three minutes, as a function of the relative optimization parameters f_T and f_p. $\overline{u}$ was taken from nearby tower measurements in height of 20 m. The curves denote borders of the optimization region. The inset shows a histogram of the errors for parameter pairs in the optimized region.

The optimization conditions Eqs. 4.16 and 4.17 can then be summarized as

$$f_T \leq f_p , \qquad [4.20a]$$

$$f_T \geq 1/f_p . \qquad [4.20b]$$

and the optimal parameters are reached for $f_T = f_p = 1$.[1]
For the analysis, all lidar data was cut into time slices of 3 minutes during which $\overline{u}$ can be seen as constant. Velocity estimates with low SNR were rejected. The variability in each grid cell was computed as the mean absolute velocity change from one sweep to the next, the grid cell variability of the complete 3 minute time slice is the average over the

---

[1] Eq. 4.20a is valid only when $\sqrt{C_s\overline{u}} \leq \Delta p_{min}$. Otherwise, it should read $f_T \leq f_p \cdot \Delta p_{opt}^2/(C_s\overline{u})$.

results in all grid cells. The final results are shown in Fig. 4.7. An inset in the figure shows an error histogram for all data inside the optimization region (i.e., data points with $f_p \geq 1$ which lie in between the two curves). In general, the computed errors show the same qualitative behavior as those retrieved from lidar simulations, and it is clearly visible that optimizing the scan parameters leads to the desired error reductions. For the present lidar, an average time undersampling error in the optimized region would be about $0.25 \ m/s$.

In application, the real time undersampling error can always be computed from the present data, and overview plots like Fig. 4.7 can help in adjusting an optimized scan pattern to the desired error limits.

For highly synchronized dual-Doppler systems, data from one lidar scan are sufficient for the retrieval, which means halving T_0 and β in Eqs. 4.16 and 4.17 (cf. Chap. 3.3). We see from Eq. 4.18, that this leads to a reduction of the optimized range gate length and time constant each by a factor $\sqrt{2}$, and therefore a much better spatial and temporal resolution in the realms of $\Delta p_\mathrm{opt} > \Delta p_\mathrm{min}$. For such applications, a synchronous control system as described in Chap. 3.1.2 is necessary.

5. The Virtual Dual-Doppler Lidar Data Set

In this chapter, the large-eddy simulation data sets are described which are the basis for the virtual dual-Doppler lidar measurements, including the boundary conditions used in the PALM model (cf. Chap. 3.2.2). Furthermore, the parameters for simulation and retrieval of the dual lidar data are discussed, followed by a quality analysis of the virtual measurements based on the error discussion in Chap. 4.
The LES were provided by Dr. Christoph Knigge and Carolin Helmke, PALM group, Institute of Meteorology and Climatology at Leibniz Universität Hannover.

5.1. Large-Eddy Simulation Data

The analyses of Chaps. 7 and 6 are based on four 30 min data sets of LES data, generated by the PALM model. The main difference between the four data sets is the pre-defined geostrophic wind of {0 m/s, 5 m/s, 10 m/s, 15 m/s} in x-direction, respectively. All simulations were computed on a grid of 5 km length in each direction, with a resolution of $\Delta x = \Delta y = \Delta z = 10$ m below $z = 1800$ m, and a slightly larger Δz for higher z. The time resolution was 1 s. All simulations were carried out for dry atmospheres over a flat surface and driven by a constant kinematic heat flux $\overline{w'\theta'}_s$ at the surface and the geostrophic wind. In all

Variable	Bottom Boundary	Top Boundary
Pressure	$\frac{\partial p}{\partial z} = 0$ hPa/m	$p = 0$ hPa
Horizontal Wind	$u = v = 0$ m/s	$u = u_G,\ v = v_G$
Vertical Wind	$w = 0$ m/s	$w = 0$ m/s
Potential Temperature	$\frac{\partial \theta}{\partial z} = 0$ K/m	$\frac{\partial \theta}{\partial z} = \frac{\partial \theta}{\partial z}_{\text{initial}}$
TKE	$\frac{\partial e}{\partial z} = 0$ m/s^2	$\frac{\partial e}{\partial z} = 0$ m/s^2

Table 5.1.: Top and bottom boundary conditions of atmospheric variables in the simulated boundary layers.

simulations with background wind the kinematic surface heat flux was 0.03 K m/s. In the calm situation it was set to 0.23 K m/s to induce strong convection. All simulations started with a pre-defined vertical potential temperature profile indicating a stable boundary layer, with a lapse rate of $d\theta/dz = 0.08$ K/m below $z = 1200$ m and $d\theta/dz = 0.74$ K/m above.

The time interval of the data output followed the simulation spin-up, consisting of a 1D-model pre-run (except in the calm situation) and one hour of 3D-simulation. In this simulation random impulses of 0.25 m/s amplitude were imposed on the wind field in 100 s intervals to initiate turbulence despite the flat boundary.

The top and bottom boundary conditions for the atmospheric variables are summarized in Tab. 5.1. Laterally, periodic boundary conditions were applied for all variables. Furthermore, a Prandtl-layer was assumed between the roughness length $z_0 = 0.15$ m (corresponding to a vegetation of hedges and few trees, cf. Stull, 1988) and the first grid points at 5 m height. The integration of the governing equations was carried out using the numerical schemes described in Chap. 3.2.2.

Note that, in the LES, u is the wind field component aligned with the geostrophic wind, and v is the corresponding perpendicular compo-

u_G [m/s]	u_* [m/s]	w_* [m/s]	L_* [m]	$z_i\,(\theta)$ [m]	$z_i\,(\overline{w'\theta'})$ [m]	$-z_i/L_*$	stability
0	0.04	2.09	-0.02	1330	1208	57000	very unstable
5	0.32	0.84	-80	1142	602	7.2	very unstable
10	0.51	0.85	-330	1114	613	1.9	unstable
15	0.68	0.87	-778	1152	665	0.8	unstable

Table 5.2.: Atmospheric scaling Parameters in the LES data sets.

nent. Only after simulation and retrieval the wind fields at evaluation height were rotated in the mean wind direction for further evaluation (cf. Eq. 5.1).

The data output consisted of the three-dimensional wind field with 1 s time resolution, as well as 10 min averages of temperature and flux profiles. The boundary layer height was computed from the height at the minimum of the vertical heat flux (Deardorff et al., 1980) as well as from the bottom of the lowest temperature inversion (Kaimal et al., 1976). From the profiles, the friction velocity u_* and convective velocity w_* were computed with Eq.2.2, using the flux time and spatial averages at the lowest grid points (cf. Tab. 5.2) and $z_i(\overline{w'\theta'})$. The resulting Obukhov-length L_* (Eq. 2.3) and the stability parameter $-z_i/L_*$ indicate unstable stratification for all simulated boundary layers.[1] This result agrees with the lapse rate of the temperature profiles in Fig. 5.1. This figure also shows the profiles of the kinematic sensible heat flux and mean wind throughout the simulated boundary layers. In comparison with the stability regimes of earlier studies (cf. Chap. 2), the development of streaks can expected here for $u_G > 0$, whereas the simulation with $u_G = 0$ has ideal conditions for the development of hexagonal convective structures.

---

[1]Although it is disputable considering the profiles in Fig. 5.1 which method for boundary layer height estimation should be favored, the choice has no effect on L_* and the overall stability classification.

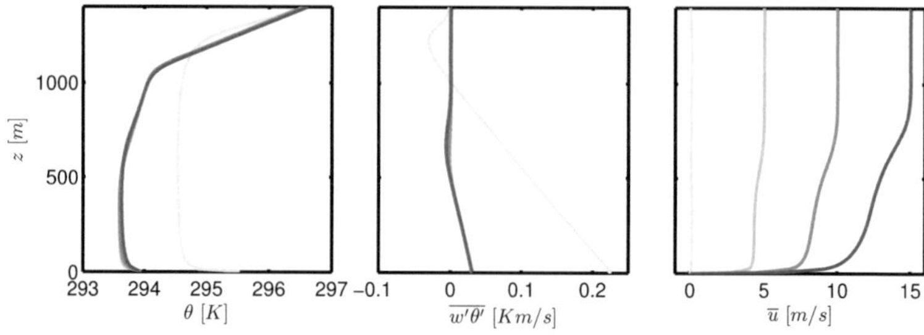

Figure 5.1.: Profiles of the potential temperature θ (left), the kinematic sensible vertical heat flux $\overline{w'\theta'}$ (center), and the mean wind $\bar{u}$ (right) for the four LES data sets (darker blue means higher u_G). $\overline{w'\theta'}$ includes the heat flux on the sub-grid scale.

To obtain reliable results from the LES independent of the subgrid-scale model, it must be ensured that the resolved-scale energy (E_{GS}) is much larger than the subgrid-scale energy (E_{SGS}). The intended virtual lidar measurements (see below) evaluate the LES at 10 m height. Fig. 5.2 shows that the average TKE profiles yield a ratio $(E_{GS} + E_{SGS})/E_{GS}$ between three and ten, which shows that a this height, the LES are only just reliable.

It should be noted that Maronga (2013) discussed that the LES is not accurate in the lowest six resolved layers. However, the goal of this work is to assess the performance of dual-Doppler lidar measurements in coherent structure detection. For this study, it is sufficient that the LES used for comparison exhibit structures comparable to those detected in real surface layers. The present LES are the best approximations of virtual boundary layers that could be obtained with manageable computational cost for simulation and retrieval.

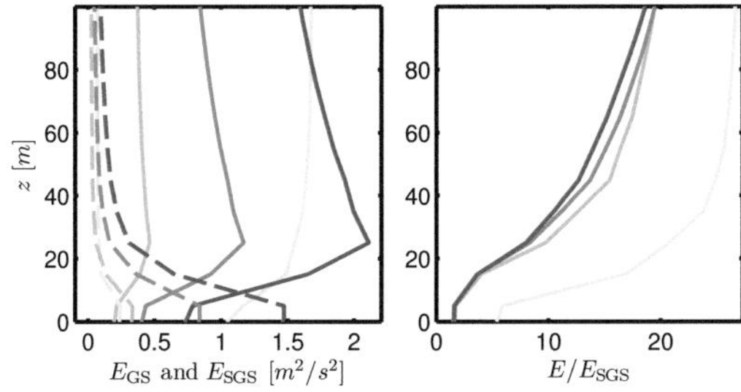

Figure 5.2.: Mean profiles of grid scale (E_{GS}) and sub-grid scale (E_{SGS}) turbulent kinetic energies in the LES data sets (left), as well as the relation of full TKE, $E = E_{\text{GS}} + E_{\text{SGS}}$, to the sub-grid scale TKE. Darker shades correspond to higher u_G.

5.2. Single Lidar Simulations

For optimal dual-Doppler retrieval results, the virtual lidar measurements were performed with the lidar simulation tool (Chap. 3.2.3) according to the error reduction technique developed in Chap. 4.2. The LES data sets covered a horizontal area of 5000 m by 5000 m, in which the lidars were positioned at (x_1, y_1, z_1) = (5000 m, 2500 m, 10 m) and (x_2, y_2, z_2) = (2500 m, 0 m, 10 m), respectively (cf. Fig. 5.3). Each lidar scanned at a constant elevation of $0°$. The azimuth sectors spanned $90°$ each, i.e. $az_1 = 315° - 45°$ and $az_2 = 225° - 315°$. Fig. 5.3 shows that the overlap region is chosen such that the lidar beams are almost perpendicular in the center and the error multiplying pre-factor (Eq. 4.9 and Fig. 4.3) is thus relatively small.

For realistic results, the lidar parameters were, wherever possible, chosen as those typical of the KIT Doppler lidars: measurements are performed with a frequency of 10 Hz with 110 range gates, starting at an offset distance of 350 m from the lidars and up to a maximal distance of 5520 m to cover the full data area. The pulse width was set

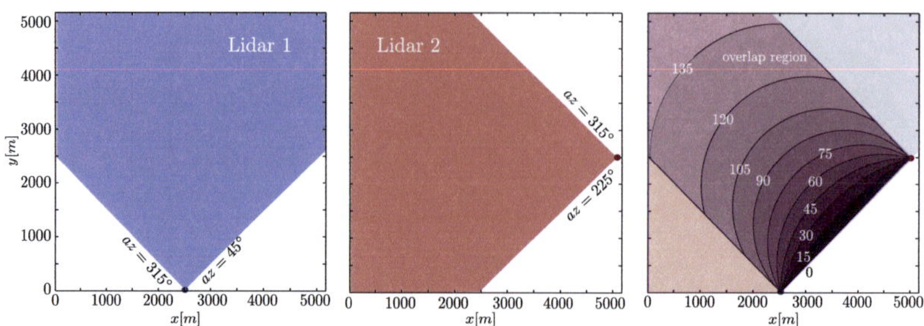

Figure 5.3.: Horizontal scanning areas on the LES data sets for the first (left) and second (center) virtual lidar. The right panel shows the overlap region with contour lines at constant lidar beam intersection angles $\Delta\chi\,[°]$.

to 300 nm and 370 nm for Lidar 1 and Lidar 2, respectively. The minimal range gate length in the optimization algorithm (Eqs. 4.18) is set to 60 m, which is slightly smaller than the pulse width, to allow for the highest possible resolution. From these parameters and the wind speed at measurement height, the range gate lengths Δp and the scanning time T_0 were derived via the optimization algorithm, as summarized in Tab. 5.3. The full set of simulation parameters can be found in Tab. B.1. Since the lidar scans were synchronized, i.e. the beams reach the turning points of the scan simultaneously, the beam angle $\beta = 90°$ and T_0 equaled the duration of one sweep.

u_G [m/s]	0	5	10	15
Wind Speed at 10 m Level [m/s]	0.0	3.0	5.1	6.8
Range Gate Length Δp [m]	60.0	60.0	66.0	76.8
Scan Time T_0 [s]	14.6	14.6	13.1	11.4
Angular Scan Velocity ω_0 [s^{-1}]	6.2	6.2	6.9	7.9

Table 5.3.: Optimized lidar simulation parameters for the four LES data sets.

5.3. Dual-Doppler Retrieval

The dual lidar data were retrieved using the software described in Chap. 3.3. Owing to synchronization, the retrieval was operated in 'one sweep mode', i.e. the horizontal wind field was computed for time intervals with the length of one beam sweep. The grid constant Δxy was chosen as the range gate length Δp which determines the highest achievable resolution (even though the range gate center distance was smaller). The grid cell radius of influence R was set to the smallest possible value, $R = \Delta xy/\sqrt{2}$. The retrieval parameters are summarized in Tab. 5.4.

The retrieval was performed on axes aligned with the Cartesian grid of the LES, and consequently the wind field components $\tilde{u}, \tilde{v}$ were the projections of the wind vector on the respective axes. After the retrieval, they were converted to the wind field component in mean wind direction, u_{RET}, and the associated component in crosswind direction, v_{RET}, and the fields were rotated to align the x-axis with the mean wind direction. The mean wind used for this conversion was derived from the original LES data for better comparison, with the mean wind direction and the crosswind direction unit vectors, $\mathbf{e}_x$ and $\mathbf{e}_y$, defined in the LES axes by

$$\mathbf{e}_x = \frac{1}{\sqrt{\langle\tilde{u}\rangle^2 + \langle\tilde{v}\rangle^2}} \begin{pmatrix} \langle\tilde{u}\rangle \\ \langle\tilde{v}\rangle \end{pmatrix}, \qquad [5.1a]$$

$$\mathbf{e}_y = \begin{pmatrix} 0 & -1 \\ 1 & 0 \end{pmatrix} \mathbf{e}_x, \qquad [5.1b]$$

where $\tilde{u}, \tilde{v}$ are the time and spatial means of the original horizontal LES wind field components at height 10 m.

Parameter	$u_G = 0$ m/s	$u_G = 5$ m/s	$u_G = 10$ m/s	$u_G = 15$ m/s
Time Constant T_0 [s]	14.6	14.6	13.1	11.4
Grid Resolution Δxy [m]	60.0	60.0	66.0	76.8
Grid Cell Radius R [m]	42.4	42.4	46.6	54.3

Table 5.4.: Retrieval parameters for the virtual dual lidar measurements in the four LES data sets.

5.4. LES Data Sets for Comparison

The original LES data are stored on a staggered grid (Fig. 3.4), where u and v are given at 5 m and 15 m height and w at 10 m height. For the comparative analysis in Chaps. 7 and 6, u and v had to be interpolated to the measurement height of 10 m using cubic splines (Bronstein et al., 2001). Even though this method implies a certain amount of smoothing, it outperforms linear or nearest-neighbor interpolation in capturing the strong curvature of the wind profile this close to the ground. Additionally, u and v were interpolated to the x- and y-axes of w to achieve an evaluation at the same grid points. Subsequently, the fields were converted into streamwise and spanwise component. Additionally, they were rotated in mean wind direction in the same manner as the retrieval fields and using the same $\mathbf{e}_x$ and $\mathbf{e}_y$ (Eq. 5.1). All comparative LES results in the following chapters were produced from these interpolated and rotated fields, u_LES and v_LES, unless specifically stated otherwise.

To estimate the influence of the temporal averaging process involved in the dual-lidar simulation and retrieval, the interpolated horizontal LES data were averaged over the retrieval time intervals T_0, resulting in a data set with full spatial LES resolution and a time resolution of the retrieval. This data set is called the time averaged LES data set.

All analyses in the following chapters are applied to the retrieval data (abbreviated as RET), the high-resolution LES data (LES) and the time-averaged LES data (LESAVG).

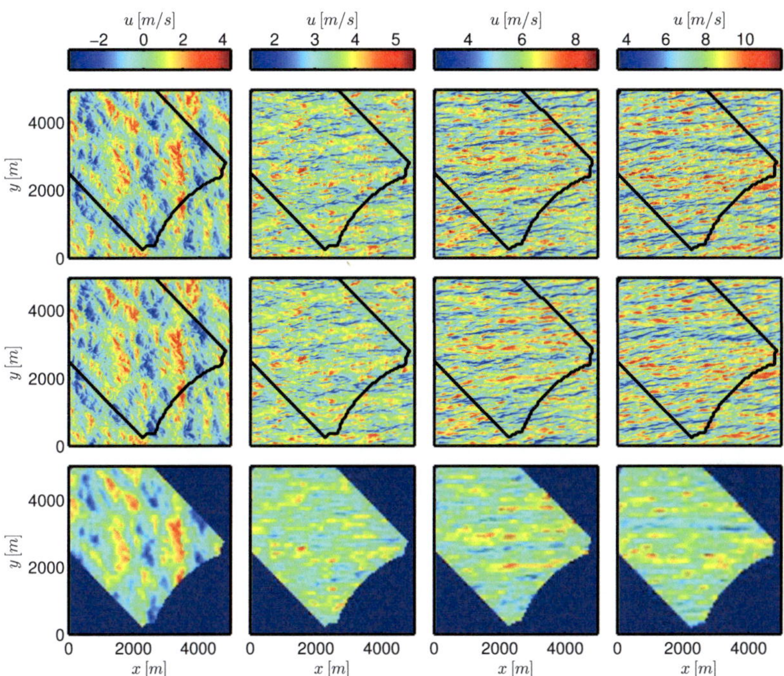

Figure 5.4.: Comparison of the wind-parallel field u from LES (top), LESAVG (center) and RET (bottom) for $u_G = \{0, 5, 10, 15\}$ m/s (left to right) at a random retrieval time step. The LES data shown correspond to the center of the retrieval time interval. The axes are the original LES axes without rotation in the mean wind direction. Note the difference in color scale between the columns.

A comparison between the three data sets is shown in Fig. 5.4. The time-averaging has only a small effect on the visible turbulence structures, whereas the dual-lidar retrieval data show considerable spatial smoothing and a much less structured field.

5.5. Quality of the Horizontal Wind Field Retrieval

The errors in the virtual lidar measurements can be analyzed in the framework presented in Chap. 4. The lidar simulator exhibits neither

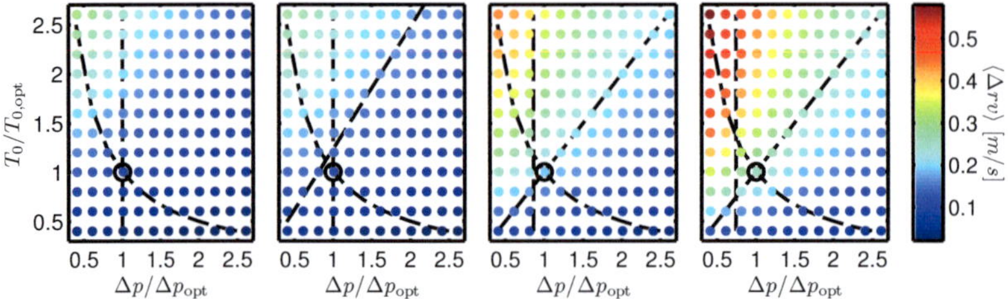

Figure 5.5.: Virtual lidar time undersampling error as a function of range gate length Δp and sweep time T_0 (relative to their optimized values) as derived from one back-and-forth sweep of Lidar 1 for $u_G = \{0, 5, 10, 15\}$ m/s (left to right). $\langle \Delta rv \rangle$ is the absolute change of mean radial velocities in a grid cell from one sweep to the next, averaged over all grid cells. The black lines denote the optimization boundaries and Δp_{min}. Optimized values are circled. Cf. Fig. 4.7 for real lidar data.

instrumental noise nor biases in the angles or the velocity estimator. However, certain errors in the velocity estimation occur since the LES grid data has to be interpolated to the virtual lidar beam. This error can be treated as a random error on the velocity estimate, $\sigma_i^{rv,rnd}$. All other single lidar errors, as listed in Tab. 4.1, are zero. Thus the single lidar propagated error results solely from $\sigma_i^{rv,rnd}$. To limit the influence of this random noise, the retrieval area was reduced to include only beam intersection angles $\Delta\chi$ beween 30° and 150°, thus limiting the error magnifying factor to $|\sin(\Delta\chi)|^{-1} \leq 2$ (Eq. 4.9 and Fig. 4.3).

Spatial and time averaging errors are as relevant in virtual lidar data as they are in real measurements. The time averaging error was minimized by the optimized scan, however, the required data density limits the achievable accuracy. Its magnitude can be estimated from Fig. 5.5, where the time sampling error was computed for the present LES data sets around their respective optimization point: the change of mean radial velocity in a grid cell from one sweep to the next, averaged over all grid cells, is $\langle rv \rangle = \{0.10, 0.12, 0.19, 0.25\}$ m/s for $u_G = \{0, 5\,10, 15\}$ m/s. Note that for $u_G = 0$ m/s and $u_G = 5$ m/s the minimal range gate length

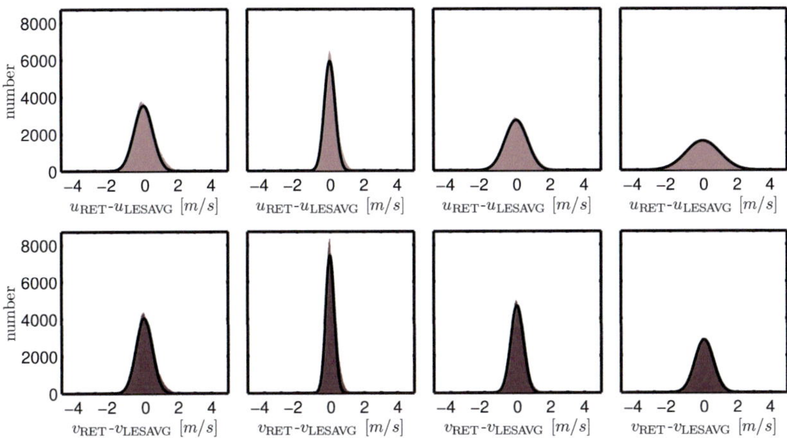

Figure 5.6.: Distribution of discrepancies between the horizontal wind fields components from the retrieval and the time averaged LES (interpolated to retrieval axes).

determines the optimization, for $u_G = 0$ m/s the error limit curve is too steep to even be displayed. In general, the time undersampling errors exhibit the expected variability with the range gate length and the time constant: the error decreases with increasing Δp and increases with increasing T_0.

Fig. 5.6 shows a comparison between the retrieval and the time-averaged LES wind fields. For the comparison, the time-averaged LES fields were interpolated linearly to the retrieval axes.

All errors are well described by a Gaussian distribution, the fit results are shown in Tab. 5.5. As expected, the standard deviation of u increases with the mean wind speed. The increase is smaller in the crosswind component due to the shifting wind direction and increased small scale shear turbulence. The larger surface heat flux in the calm situation has the same effect of increasing small scale turbulence. In comparison,

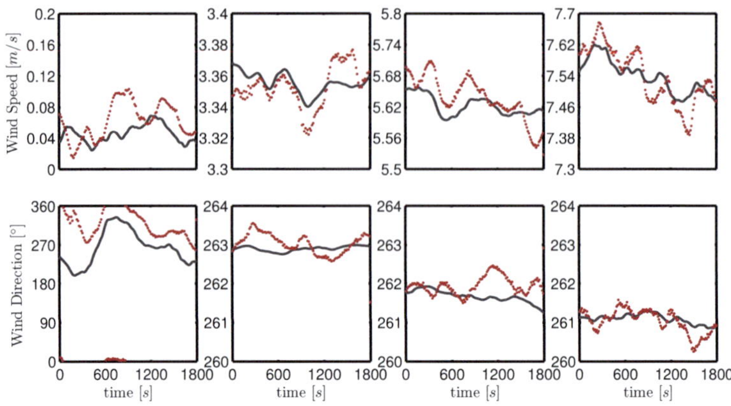

Figure 5.7.: Comparison of the horizontal means of wind speed (top) and wind direction (bottom) as obtained from the retrieval data (red) and the LES data (black) for all four LES data sets (u_G increasing from left to right).

the bias is for both components on the scale of few cm/s and therefore negligible.

Naturally, the difference between the spatial means of the retrieval and LES data are smaller than the errors in the point-by-point comparison above. Fig. 5.7 shows that the spatial mean wind is captured by the retrieval with errors smaller than 0.1 m/s. The wind direction, in the presence of background wind, exhibits errors smaller then 1°. However, the strong differences in the absence of geostrophic wind are a result of the negligible absolute values.

	0 m/s		5 m/s		10 m/s		15 m/s	
Wind Field	$\sigma\left[\frac{m}{s}\right]$	$\mu\left[\frac{m}{s}\right]$	$\sigma\left[\frac{m}{s}\right]$	$\mu\left[\frac{m}{s}\right]$	$\sigma\left[\frac{m}{s}\right]$	$\mu\left[\frac{m}{s}\right]$	$\sigma\left[\frac{m}{s}\right]$	$\mu\left[\frac{m}{s}\right]$
u	0.56	-0.08	0.33	-0.04	0.68	-0.05	1.02	-0.06
v	0.49	-0.07	0.26	-0.04	0.39	-0.02	0.56	-0.00

Table 5.5.: Standard deviations σ and means μ as obtained by a least-squares fit of a Gaussian function to the distributions in Fig. 5.6

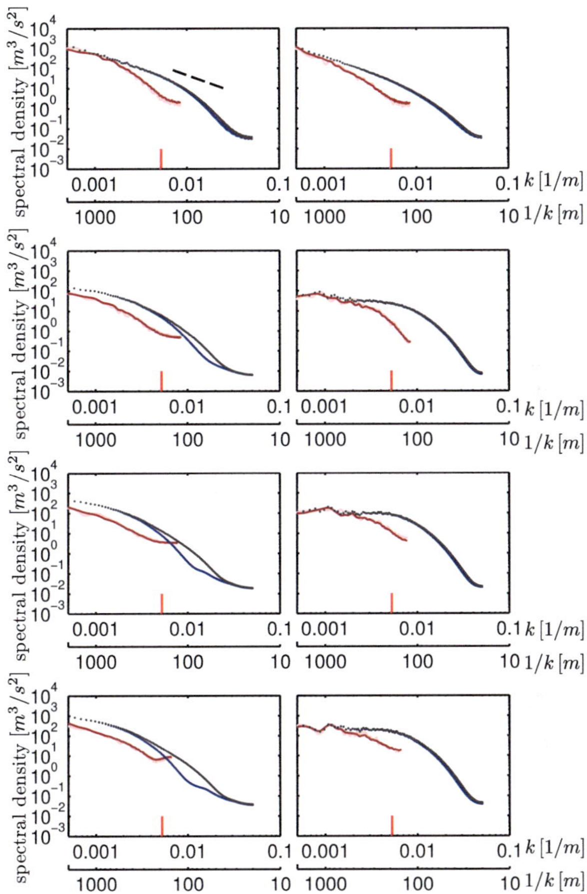

Figure 5.8.: Comparative spectral density of the u-component along (left) and across (right) the mean wind direction for the four data sets (u_G increasing from top to bottom). The spectra are shown for the retrieval results (red), the time-averaged LES results (blue) and the LES results (black). The dashed line indicates the slope of $k^{-5/3}$. The pale red lines show a random choice of ten retrieval spectra. The red mark on the k-axis indicates the effective resolution of the simulation and retrieval.

Fig. 5.8 shows the comparative spectral density of the u-component of the retrieval data, the time-averaged LES data and the original LES data. The spectra were computed from all time steps of the retrieval and the time-averaged LES, as well as from 100 random time steps in the LES. For better resolution, the wind fields at each time step were zero-padded to 1024 grid points before performing the Fourier transform. The LES-spectra agree well with the time-averaged LES results, with deviations only for high wave numbers k and, caused by advection, increasingly for higher wind speeds. Both show approximately the expected decay with $k^{-5/3}$ in the inertial range (Stull, 1988). The slight increase around the highest wave number can result from interpolation effects. This is also visible in the retrieval data.

It is obvious that the virtual lidar results fail to resolve the full spectral energy already on scales much larger than the effective retrieval scale (cf. Chap. 3). For small wave numbers, the agreement is very accurate for the calm situation and the spectra in y-direction. However, with increasing background geostrophic wind the scales up to which the spectra deviate becomes considerably larger, up to half an order of magnitude loss of spectral density at $1/k = 1$ km for $u_G = 15$ m/s. This effect is visible neither in the y-spectra of the u nor in the v-component (cf. the v-spectra in App. C), and only very slight in the x-spectra of v. The sole occurrence of the large-scale spectral underestimation for x-spectra, its dependency on the background wind, and the stronger deviation in the streamwise component indicate that this effect is caused by advection.

The effect is investigated in more detail in Fig. 5.9, where the time-averaged LES spectra are reproduced alongside spectra of the time-averaged LES wind fields which were spatially smoothed in x-direction, y-direction and both x- and y-direction using a moving average.

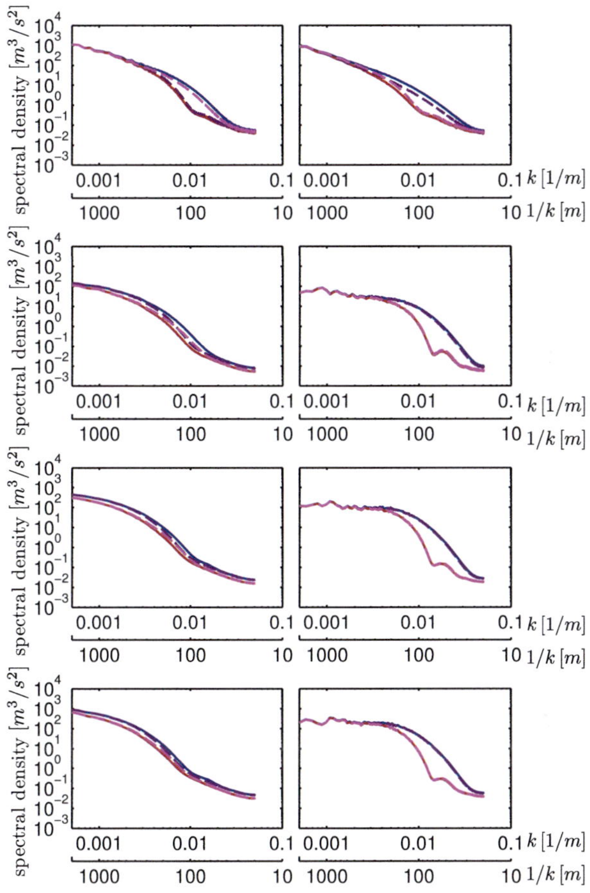

Figure 5.9.: Effect of spatial smoothing on the spectral density of the u-component along (left) and across (right) the mean wind direction for the four data sets (u_G increasing from top to bottom). The mean spectra are shown for the time-averaged LES results after applying a moving average filter with the span (Δ_x, Δ_y) in x- and y-direction, respectively: $(\Delta_x, \Delta_y) = (0\text{ m}, 0\text{ m})$ (blue), $(\Delta_x, \Delta_y) = (\Delta, 0\text{ m})$ (dark purple), $(\Delta_x, \Delta_y) = (0\text{ m}, \Delta)$ (light purple), and $(\Delta_x, \Delta_y) = (\Delta, \Delta)$ (red). $\Delta = \{70\text{ m}, 70\text{ m}, 70\text{ m}, 90\text{ m}\}$ for $u_G = \{0\text{ m/s}, 5\text{ m/s}, 10\text{ m/s}, 15\text{ m/s}\}$.

The corresponding results for the v-component are shown in App. D. Averaging over lengths Δ in the direction of spectral analysis leads, as expected, to a loss in spectral density at high wave numbers k, i.e. at $k \geq 1/(2\Delta)$. Furthermore, it becomes obvious that the spatial averaging across the direction of spectral analysis has a substantial influence as well: it leads to a decrease in spectral density which is approximately constant over the full range of wave numbers. As in Fig. 5.8, this underprediction becomes apparent only when the direction of spectral evaluation coincides with the mean wind direction, and it increases with u_G. Averaging in both directions yields both types of loss-effects.

Qualitatively, this result can be expected when streaky structures are energetically dominant: when the structure widths becomes smaller than Δ, the averaging process in y-direction levels the wind field significantly, thereby reducing the spectral content in x-direction as well. On the other hand, when the streak-length in x-direction is larger than Δ, the x-directional averaging will not have a large effect on the y-spectra. Therefore, the large-scale spectral loss is caused by advection in the sense that shear shapes the streaky structure of the surface layer wind fields.

6. Assessment of Dual-Doppler Lidar Capability to Detect and Quantify Aspects of Coherent Structures

This chapter presents the results from comparative coherent structure length scale analysis on the LES and the virtual lidar data. Three methods were used for coherent structure detection: measurement of spatial integral length scales, wavelet analysis, and structure clustering. The influence of the lidar averaging processes is furthermore investigated theoretically, leading to quality control and correction techniques for the methods.

6.1. Spatial Correlation and Integral Length Scales

Integral length scales are a common tool to investigate the length scale up to which fields are correlated (cf. Chap. 2). In this way, the streamwise elongation of streaks and the resulting anisotropy can be analyzed. With this method only the mean correlation length can be derived, the single structures and their positions remain unknown.

6.1.1. Correlation Length Definitions

The autocorrelation function of a scalar field f at lag x is defined as

$$r_f(x) = \frac{1}{\sigma_f^2} \langle f'(x+x')\,f'(x') \rangle_{x'}, \qquad [6.1]$$

where

$$f' = f - \langle f \rangle \qquad [6.2]$$

is the turbulent component and

$$\sigma_f^2 = \langle f'^2 \rangle \qquad [6.3]$$

is the variance.

Here, $\langle \cdot \rangle$ denotes the average over all arguments in the function, typically space and time variables, and $\langle \cdot \rangle_x$ means an average over the argument x of the function.

The spatial autocovariance $\rho_f(x)$ is defined as the non-normalized auto-correlation:

$$\rho_f(x) = \sigma_f^2 r_f(x) . \qquad [6.4]$$

For fields with a finite number of data points $f = \{f_1, f_2, \ldots, f_N\}$, the definition becomes

$$r_{f,i} = \frac{1}{\sigma_f^2} \frac{1}{N-i} \sum_{k=1}^{N-i} f'_{i+k} f'_k, \quad i = 0, 1, \ldots, N-1 , \qquad [6.5]$$

where i is the relative shift.

Accordingly,

$$\sigma_f^2 = \frac{1}{N} \sum_{j=1}^{N} f_j'^2 . \qquad [6.6]$$

If f is a scalar function on the 2D plane, the autocorrelation function can be computed in two dimensions,

$$r_f(\mathbf{x}) = \frac{1}{\sigma_f^2} \langle f'(\mathbf{x} + \mathbf{x}') f'(\mathbf{x}) \rangle_{\mathbf{x}'} , \qquad [6.7]$$

where $\mathbf{x}$ is a vector in the plane.

For discrete fields $f = \{f_{(1,1)}, f_{(1,2)}, f_{(2,1)} \cdots, f_{(N_1,N_2)}\}$,

$$r_{f,(i,j)} = \frac{1}{\sigma_f^2} \frac{1}{(N_1 - i)(N_2 - j)} \sum_{k=0}^{N_1 - i} \sum_{l=0}^{N_2 - j} f'_{(i+k,j+l)} f'_{(k,l)} , \qquad [6.8]$$

with

$$\sigma_f^2 = \frac{1}{N_1 N_2} \sum_k \sum_l f_{(k,l)}'^2 \,.$$

[6.9]

These definitions can easily be extended to three spatial dimensions and the time dimension.

The integral scales are length scales which measure the distance up to which the wind field can be seen as correlated with itself. The integral scale for a function f on $\mathbb{R}^1$ is defined as

$$L_f = \int_0^\infty dx\, r_f(x) \,,$$

[6.10]

where r_f is the spatial autocorrelation a defined above. A common approximation (Lenschow and Stankov, 1986) is to integrate only up to the first zero-crossing of the autocorrelation:

$$L_f \approx \int_0^{r \equiv 0} dx\, r_f(x) \,,$$

[6.11]

which is to the first maximum of the integral. This approximation is used here since the computed autocorrelation becomes increasingly noisy for higher lags x.

For functions f on $\mathbb{R}^2$, the integral scale is furthermore a function of the direction in which the correlation is analyzed. For the horizontal plane, the x-axis is defined as usual as the direction of the mean wind, and the y-axis as the right-handed axis perpendicular to the x-axis. $L_{f,x}$ and $L_{f,y}$

are defined as the integral length scales in x- and y-direction, respectively. Using Eq. 6.11,

$$L_{f,x} = \int_0^{r\equiv0} dx\, r_f(x \cdot \mathbf{e}_x) \qquad [6.12\text{a}]$$

$$\text{and} \quad L_{f,y} = \int_0^{r\equiv0} dy\, r_f(y \cdot \mathbf{e}_y) \,. \qquad [6.12\text{b}]$$

For discrete wind fields, the spacings Δx and Δy between adjacent components of f must be taken into account:

$$L_{f,x} = \sum_{i=0}^{r\equiv0} r_{f,(i,0)}\, \Delta x \,, \qquad [6.13\text{a}]$$

$$L_{f,} = \sum_{j=0}^{r\equiv0} r_{f,(0,j)}\, \Delta y \,. \qquad [6.13\text{b}]$$

The anisotropy or aspect-ratio of the wind field can be measured using

$$A_f = \frac{L_{f,x}}{L_{f,y}} \,. \qquad [6.14]$$

6.1.2. Theoretical Considerations

Integral length scales from dual-lidar data

A theoretical prediction about the ability of the lidar to estimate correlation lengths accurately can be derived from the averaging processes that influence the dual-lidar retrieval data:

From the mathematical models for single lidar velocity estimation (Eqs. 3.5-3.8) and the dual-lidar retrieval techniques (Chap. 3.3) it is known that the dual-lidar retrieved wind field at a certain grid point can be described as a weighted average of the real wind field in the vicinity of the grid point. Frehlich (1997) investigated the effect of single-lidar pulse averaging on velocity variance measurements and developed a method to correct the results using the single lidar weighting function. However, no analytic expression for the weighting function can be given in the dual-Doppler case, since not only spatial parameters, but also the time shift between the two scanning lidars and the wind field evolution during this time shift are factored into the final result. However, it can be safely assumed that this weighting function is approximately constant on the grid cell, zero for points far away from the grid cell and decreasing noticeably around distances from the grid cell center of the order of the lidar resolution.

This view is supported by the comparison of LES and retrieval spectra (cf. Fig. 5.8): A drop-off in spectral energy occurs in the lidar spectra as compared to the LES spectra at scales of three to four times the estimated lidar resolution. The smaller the scales become, the less spectral energy can be 'seen' by the lidar. However, on larger scales, the instrument is mostly able to resolve the full spectral content of the LES, the cases where an underestimation occurs on large scales as well can be explained by the cross-directional averaging (Fig. 5.9).

Let f be one of the fully resolved wind field components, and $\tilde{f}$ the same component as retrieved from the dual-lidar measurement. Under the assumption that the weighting is the same for all points in the plane, the retrieved field can be written

$$\tilde{f}(\mathbf{x}) = \int d\mathbf{x}' \, f(\mathbf{x}')w(\mathbf{x} - \mathbf{x}') \,, \qquad [6.15]$$

where w stands for the appropriate weighting function. Using Eq. 6.15, the autocorrelation function $r_{\tilde{f}}$ for the dual-lidar wind fields yields

$$r_{\tilde{f}}(\mathbf{x}) = \frac{1}{\sigma_{\tilde{f}}^2} \int d\mathbf{x}' \, \tilde{f}'(\mathbf{x}+\mathbf{x}')\tilde{f}'(\mathbf{x}') \qquad [6.16a]$$

$$= \int d\mathbf{x}' \, r_f(\mathbf{x}+\mathbf{x}')W(\mathbf{x}') , \qquad [6.16b]$$

with

$$W(\mathbf{x}) = \frac{\sigma_f^2}{\sigma_{\tilde{f}}^2} \int d\mathbf{x}' \, w(\mathbf{x}')w(\mathbf{x}'+\mathbf{x}) . \qquad [6.17]$$

This means that the lidar autocorrelation function is a smoothed version of the fully resolved autocorrelation. With our assumptions about the lidar spatial averaging function, w, the autocorrelation smoothing should occur on scales of the order of the scales in w. Note that W is not normalized, so that on average $r_{\tilde{f}}$ is larger than r_f by a factor $\sigma_f^2/\sigma_{\tilde{f}}^2$. Autocorrelations r which decrease slowly can appear almost linear to the smoothing function, so the autocorrelation functions from the LES and from the lidar data should nearly coincide. On the other hand, if the fully resolved autocorrelation decreases rapidly to zero, the positive curvature is noticeable in the smoothing. The effect on the correlation length can be estimated using the example of an exponential autocorrelation function (Lothon et al., 2006): Let $r(x) = \exp(-|x|/L)$ represent the fully resolved 1D-autocorrelation with integral scale L, and let w be given by a constant average over a $2x_0$ interval (cf. Fig. 6.1):

$$w_{x_0}(x) = \begin{cases} \frac{1}{2x_0} , & -x_0 \leq x \leq x_0 \\ 0 , & \text{otherwise} \end{cases} . \qquad [6.18]$$

The smoothed autocorrelation can be computed from Eq. 6.16b, from which the integral scale $\tilde{L}$ is derived using Eq. 6.10:

$$\frac{\tilde{L}}{L} = \frac{\sigma_f^2}{\sigma_{\tilde{f}}^2} = \frac{\frac{1}{2}\left(\frac{2x_0}{L}\right)^2}{\frac{2x_0}{L} + \exp(-\frac{2x_0}{L}) - 1} . \qquad [6.19]$$

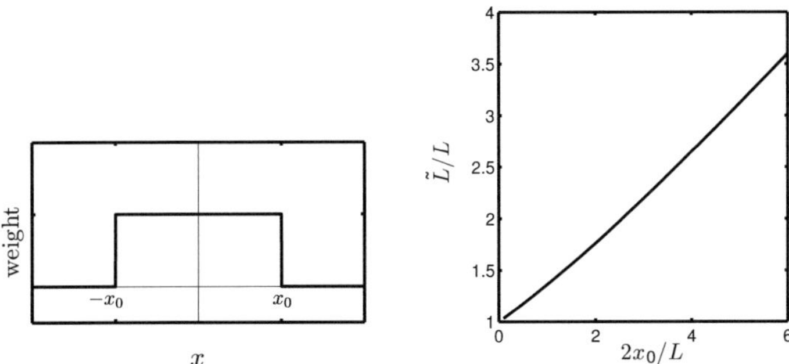

Figure 6.1.: Left: The one-dimensional weighting function w_{x_0}, Eq. 6.18. Right: The overestimation factor $\tilde{L}/L$ from Eq. 6.19.

This relation, shown in Fig. 6.1 predicts the overestimation simply as the relation of variances, which in turn is a function of the ratio between averaging scale and integral length scale. As a result, the lidar should overestimate the correlation lengths if they are of the order of or shorter than the lidar averaging scales. The overestimation effect should increase the shorter the correlation lengths become. Correlation lengths much larger than the lidar averaging scales should be accurately estimated by the lidar. If the 1D-model approach (Eq. 6.19) is valid, the overestimation factor can be computed as the relation of LES and retrieval variances.

Integral length scales from tower data

As noted in the introduction, coherent structure detection is usually based on time series from tower measurements. It is therefore necessary to discuss the accuracy of correlation length computations from point measurements.

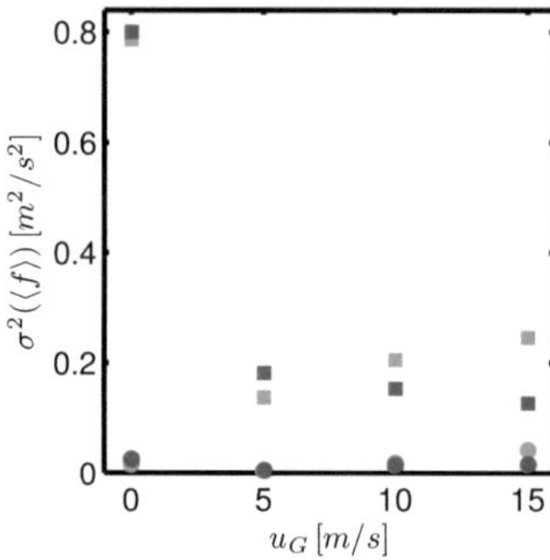

Figure 6.2.: Standard deviation of mean values of the u (light) and v (dark) fields used for the correlation length computation. The squares and circles mark the results for temporal and spatial correlations, respectively, i.e., $\sigma^2_{\langle f(\mathbf{x}',t')\rangle_{t'}}$ and $\sigma^2_{\langle f(\mathbf{x}',t')\rangle_{\mathbf{x}'}}$. Cf. Eq. 6.23

Using Taylor's hypothesis, the one-dimensional autocorrelation function of the field f in mean wind direction is given by

$$r_{f,\text{Taylor}}(x) = \frac{1}{\sigma_f^2} \left\langle f'\left(\mathbf{x}',t'\right) f'\left(\mathbf{x}',t' - \frac{x}{\langle u\rangle}\right)\right\rangle_{t'} \qquad [6.20]$$

for a tower positioned at $\mathbf{x}'$, with the mean and variance of f computed using $\langle\cdot\rangle_{t'}$.

It is important to note that, in the present LES data set, the ergodic condition (Stull, 1988) is not fulfilled: the temporal 30 min means vary considerably depending on the position of the virtual tower, and the standard deviation of the temporal means is about one order of magnitude larger than the temporal variation of the spatial means (cf. Fig. 6.2). This variability has a crucial effect on the autocorrelation computation and, as a result, on the integral scales:

Let $\rho_f(t)$ denote the temporal autocovariance of a function $f(\mathbf{x},t)$ on the time interval $[0,T]$ with time-shift t, computed from the full data set (i.e., using averages $\langle \cdot \rangle$), and let $\tilde{\rho}_f(t)$ be the average of all time series auto-covariances (Eq. 6.20), measured at fixed points $\mathbf{x}'$:

$$\rho_f(t) = \frac{1}{N_{\mathbf{x}} N_t} \sum_{\mathbf{x}',t'} \left(f(\mathbf{x}',t'+t) - \langle f(\mathbf{x}',t') \rangle \right) \left(f(\mathbf{x}',t') - \langle f(\mathbf{x}',t') \rangle \right) , \quad [6.21a]$$

$$\tilde{\rho}_f(t) = \frac{1}{N_{\mathbf{x}} N_t} \sum_{\mathbf{x}',t'} \left(f(\mathbf{x}',t'+t) - \langle f(\mathbf{x}',t') \rangle_{t'} \right) \left(f(\mathbf{x}',t') - \langle f(\mathbf{x}',t') \rangle_{t'} \right) ,$$

$$[6.21b]$$

where $N_{\mathbf{x}}$ is the number of spatial grid points and N_t the number of time steps in the overlap time interval $(T - t)$. If the temporal shift t is small compared to the full length of the time series,

$$\frac{1}{N_t} \sum_{t'} f(\mathbf{x}',t'+t) \approx \frac{1}{N_t} \sum_{t'} f(\mathbf{x}',t') \approx \langle f(\mathbf{x}',t') \rangle_{t'} , \quad [6.22]$$

which yields

$$\rho_f(t) - \tilde{\rho}_f(t) \approx \frac{1}{N_{\mathbf{x}}} \sum_{\mathbf{x}'} \left(\langle f(\mathbf{x}',t') \rangle_{t'} - \langle f(\mathbf{x}',t') \rangle \right)^2 = \sigma^2_{\langle f(\mathbf{x}',t') \rangle_{t'}} . \quad [6.23]$$

This shows that the averaged autocovariance from smaller subsets of the full data set is given by the full autocovariance minus the variance of the averages from the subsets. Lenschow et al. (1994) show that $\sigma^2_{\langle f(\mathbf{x}',t') \rangle_{t'}} \approx 2\rho_f(0)L_\tau/T$, if T is much larger than the temporal integral length scale L_τ. Consequently,

$$r_f(t) = \frac{\rho_f(t)}{\rho_f(0)} \approx \frac{\tilde{\rho}_f(t) + \sigma^2_{\langle f(\mathbf{x}',t') \rangle_{t'}}}{\tilde{\rho}_f(0) + \sigma^2_{\langle f(\mathbf{x}',t') \rangle_{t'}}} > \tilde{r}_f(t) \qquad \forall t : \tilde{\rho}_f(t) < \tilde{\rho}_f(0) \quad [6.24]$$

if t is sufficiently small. Therefore, the integral length scales computed from the time series will be shorter than the length scales computed from spatial data.

One can attempt to correct this by subtracting the full mean of the data set from the time series instead of the time series mean, and normalizing with the full variance instead of the series-wise variance. However, this approach may not be possible with real data, where therefore the validity of the ergodic condition has to be ascertained before comparisons between temporal and spatial data are possible.

6.1.3. Results

For u and v of all three data sets, i.e. the LES, time-averaged LES and the retrieval data, the autocorrelation function was computed using the algorithm given in Eq. 6.8 for each time step. The mean and variance in this equation were taken from the respective time step only. From the autocorrelation the integral scales were obtained for each time step using Eq. 6.13a, where multiplication with the respective grid resolution led to results in metric units. The anisotropy follows from Eq. 6.14.

Furthermore, correlation lengths were computed from 2700 time series, taken from virtual towers[1] equally distributed across the area in 100 m intervals. Time series yield only correlation length of u and v in x-direction (i.e., mean wind direction), therefore neither L_y nor the anisotropy could be computed.

Due to the known effects of the missing ergodicity in the data set (Eq. 6.24), the time series results were computed in two ways: firstly, under the assumption that the ergodic condition holds, autocorrelations were computed using the mean and variance of each time series (Eq. 6.20), and secondly, to remove errors due to fluctuations in the means of the time series, the autocorrelations were computed using the overall mean wind speed and mean spatial variance. The results from the second method will hereafter be called the corrected results.

In App. E, Tab. E.1 gives an overview of the data set size. Data loss

---

[1]The virtual towers described here, i.e. time series at fixed grid points in the LES data, should not be confused with dual-lidar virtual towers (Chap. 3.1.3)

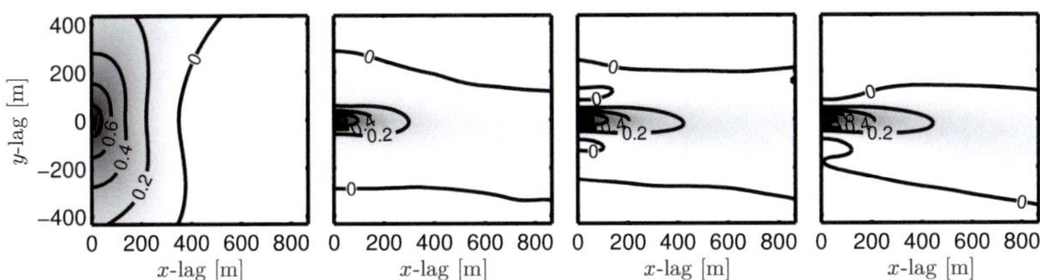

Figure 6.3.: Development of the spatial autocorrelation in the LES u wind fields with the background wind, $u_G = \{0, 5, 10, 15\}$ m/s from left to right.

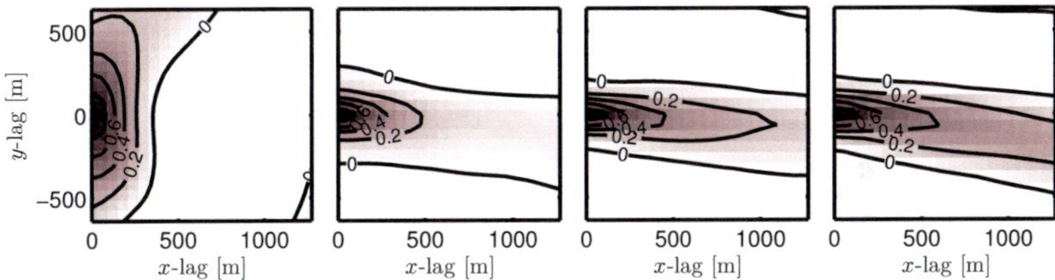

Figure 6.4.: Development of the spatial autocorrelation in the retrieved u wind fields with the background wind, $u_G = \{0, 5, 10, 15\}$ m/s from left to right.

occurred whenever the autocorrelation function had no zero-crossing in a certain direction.

App. F shows the time averages of the spatial autocorrelation for the three data sets in the $x > 0$ half plane. (Note that Eq. 6.7 shows that $r(\mathbf{x}) = r(-\mathbf{x})$, therefore, one half plane contains all information). It is evident from Figs. F.3 and F.4 that the dual lidar data are qualitatively capable of capturing the repetitive structure on the shear-driven wind fields.

The development of the u autocorrelation with increasing wind speed for the LES and retrieval data is shown in more detail in Figs. 6.3 and 6.4. The LES results (Fig. 6.3) show that, as expected, the correlation increases for lags in mean wind direction with increasing background wind, whereas the correlation in cross-wind direction y appears to decrease

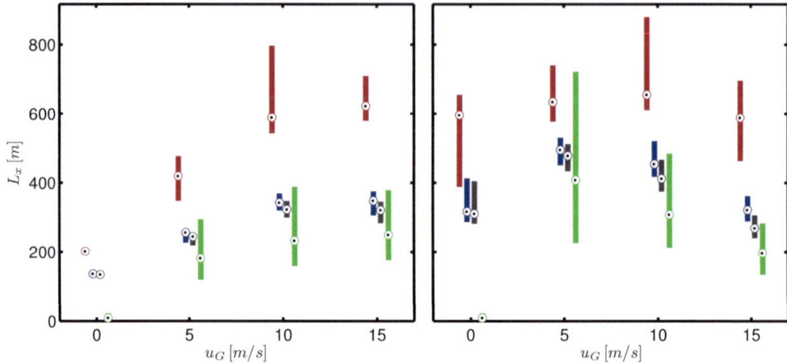

Figure 6.5.: *Integral length scales*: Scales L_x of the u (left) and v (right) wind fields in mean wind direction for all data sets: the retrieval results (red), the averaged LES fields (blue), the LES fields (black), and the time series (green). The bars cover the range between the 25th and 75th percentile, the circles mark the median.

slightly. The retrieval results (Fig. 6.4) appear qualitatively similar, however, the autocorrelation is clearly overestimated.

Fig. 6.5 shows the resulting distributions of integral correlation lengths of u and y in x- direction as box plots with the centers indicating the median and the bars ranging from 25th to 75th percentile. The figure includes the corrected results from the time series. The difference between the corrected and the uncorrected time series results is shown in Fig. 6.6. Fig. 6.7 shows the integral length scales in y-direction, and the anisotropy of both wind field components is shown in Fig. 6.8.

The LES data show that the correlation length in wind direction for u increases with wind speed, whereas it decreases with wind speed for v. Both u and v show a decrease of correlation length in y-direction. Both fields exhibit considerable anisotropy, which increases with the mean wind but appears to level at higher wind speeds. The correlation

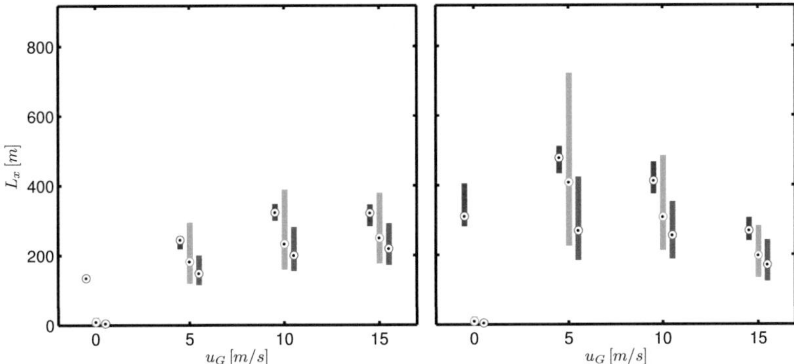

Figure 6.6.: *Integral length scales*: Scales L_x of the u (left) and v (right) wind fields in mean wind direction for the time series, computed with the full mean and variance of the data set (light green) and the mean and variance of the respective series (dark green). The black bars show the comparative LES results. The bars cover the range between the 25th and 75th percentile, the circles mark the median.

length perpendicular to the wind direction is extremely small for all wind speeds apart from zero, and approaches the lower bound, i.e. 10 m for the LES data. The time averaged LES data show nearly no difference from the LES data, albeit a noticeable slight deviation for higher wind speeds. The small overestimation of correlation lengths conforms to the theory of expected overestimation in the retrieval data when time averaging is regarded as another form of spatial averaging.

The corrected time series results in Fig. 6.5, as well as the comparison of corrected and uncorrected time series results in Fig. 6.6 behave as expected from the theoretical considerations: whereas the corrected results approximately match the LES results apart from a slight underestimation, the uncorrected results severely underestimate the integral scales although they exhibit a much more localized distribution.

The correlation lengths measured by the lidar simulator show the same qualitative development for each data set and component as the LES

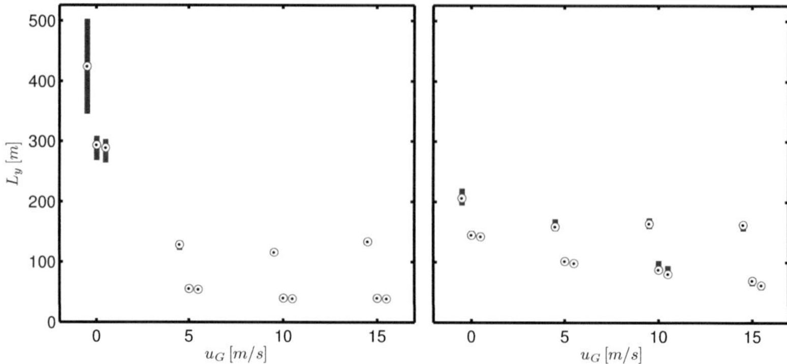

Figure 6.7.: *Integral length scales*: Scales L_y of the u (left) and v (right) wind fields in cross-wind direction for all data sets. Colors and range as in Fig. 6.5.

length scales. However, for all wind fields, the virtual lidar data exhibit a considerable overestimation of correlation lengths.

The lidar overestimation of length scales is more pronounced in the y direction, since there the correlation lengths are of the order of the lidar resolution. This leads to the effect that the lidar generally underestimates the anisotropy (cf. Fig. 6.8), an effect that becomes more pronounced the smaller the length scales in y direction become.

This was to be expected from theoretical consideration for shorter correlation lengths, but the effect should be decreasing for larger length scales.

Fig. 6.9 investigates the validity of the theoretical model (shown in the left panel): The overestimation factor is expected to approach unity if the median integral scales become several times larger than the lidar resolution. Qualitatively, the effect is clearly visible. However, the overestimation factor only goes down to $\approx$1.2, even for length scales eight times as large as the lidar resolution. This suggests that the approximate linearity of r_{LES}, which is a sufficient condition for accurate

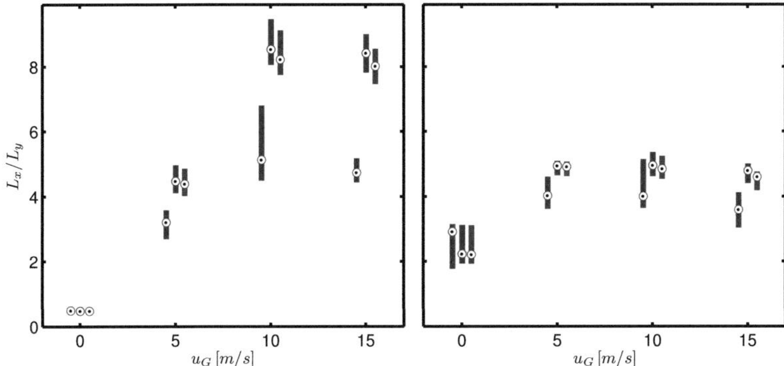

Figure 6.8.: *Integral length scales*: Anisotropy L_x/L_y of the u (left) and v (right) wind fields for all data sets. Colors and range as in Fig. 6.5.

length scale estimation, is not necessarily fulfilled even for large integral length scales. Comparing the data points with the simple theoretical model shows that the model describes the qualitative behavior well, but that the effective averaging length $2x_0$ is larger than the lidar resolution by a factor of approximately seven.

According to the 1D-model (Eq. 6.19), the overestimation factor can be predicted from the quotient of LES and lidar variances in the fields.
Fig. 6.10 shows that the equation is a good approximation. Even though variances computed from time series provide less precise results, especially for the calm situation, it is most likely that real measurement campaigns will use time series to estimate the real wind field variance. Therefore, the retrieval data shown in Fig. 6.11 were corrected by division with the overestimation factor computed from the time series variances. The factor was computed as the quotient of median variances for each wind field.

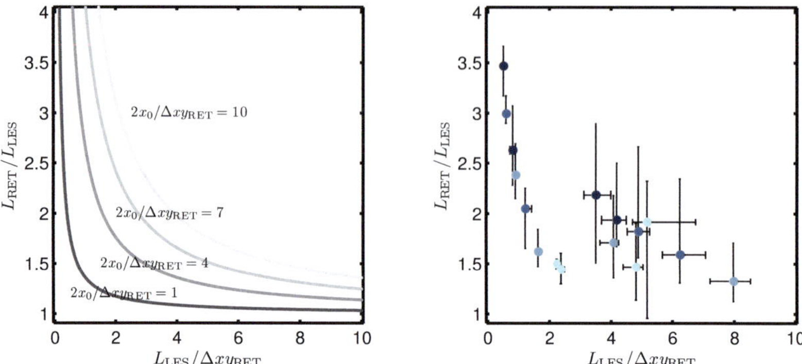

Figure 6.9.: *Integral length scales*: Overestimation factor $L_{\mathsf{RET}}/L_{\mathsf{LES}}$ as a function of $L_{\mathsf{LES}}/\Delta xy$ from theory (Eq. 6.19, left) and data (right). The theoretical curves result from Eq. 6.19 with $L = L_{\mathsf{LES}}$ and $\tilde{L} = L_{\mathsf{RET}}$ for different relations $2x_0/\Delta xy$ of lidar averaging scales to retrieval resolution. The error bars of the data results are based on the 25th and 75th percentile of the distributions, i.e. they range from $p_{25}(L_{\mathsf{RET}})/p_{75}(L_{\mathsf{LES}})$ to $p_{75}(L_{\mathsf{RET}})/p_{25}(L_{\mathsf{LES}})$ on the ordinate and from $p_{25}(L_{\mathsf{LES}})/\Delta xy$ to $p_{75}(L_{\mathsf{LES}})/\Delta xy$ on the abscissa. The data points in the right panel were computed accordingly using the medians of the distributions. Their colors indicate wind speeds, darker blue means higher u_G.

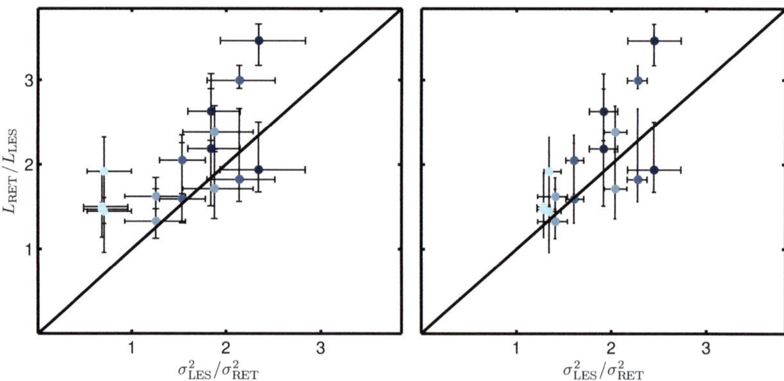

Figure 6.10.: *Integral length scales*: Overestimation factor $L_{\mathsf{RET}}/L_{\mathsf{LES}}$ for all fields $f \in \{u, v\}$ as a function of $\sigma^2_{\mathsf{LES}}/\sigma^2_{\mathsf{RET}}$ for σ^2_{LES} from virtual tower time series (left) and for σ^2_{LES} from the spatial variance in each time step (right). Data points and error bars were computed as in Fig. 6.9. The line marks the theoretical identity, cf. Eq. 6.19. The colors indicate wind speeds as in Fig. 6.9.

The correction method reduces the error notably. Except for the calm wind fields, the bias nearly vanishes. The remaining errors lead to a substantial, albeit smaller bias in the anisotropy. An important exception are the calm wind fields, where the correction increased the bias. Fig. 6.10 shows that time series in the calm situation are unable to accurately estimate the spatial variances. The correction method can therefore be applied whenever it is possible to measure spatial variances, or when ergodicity can be assumed.

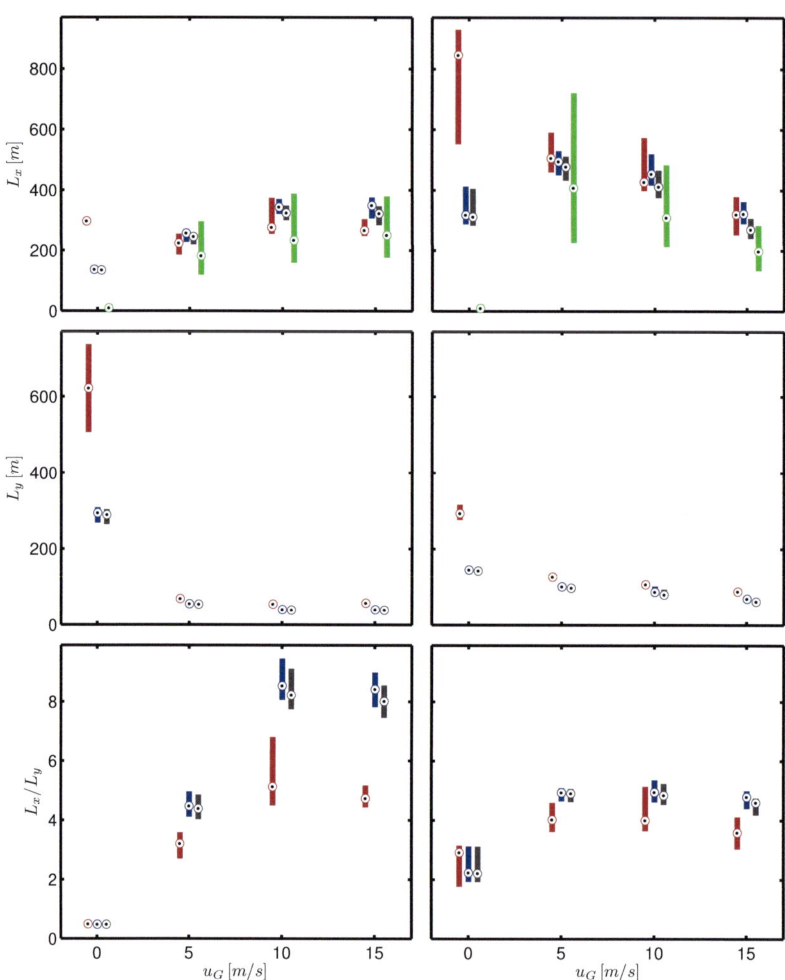

Figure 6.11.: *Integral length scales*: The results from Figs. 6.5, 6.7, and 6.8, computed with the corrected retrieval results, i.e. $L_{\text{RET,corr}} = \sigma^2_{\text{RET}}/\sigma^2_{\text{LES}} \cdot L_{\text{RET}}$, with σ^2_{LES} from virtual tower time series and σ^2_{RET} from the mean spatial variance in the respective fields.

6.2. Wavelet Analysis

Throughout the last years, wavelet analysis has become a common tool for coherent structure detection in time series data (cf. Chap. 2). The analyses differ in certain aspects, e.g. in the wavelets used for structure detection, the determination of structure lengths and the type of time series used. What they have in common is the application of the wavelet transform to a time series of data and the analysis of the transforms with respect to maxima, minima, zero crossings, and energy distributions to detect signatures of structures.

Wavelet analysis allows to detect single structures in the wind field instead of only characterizing mean length scales. Usually, only the energetically dominant structures are investigated, which leads to a certain amount of smoothing.

6.2.1. The Wavelet Transform

Dirac-Notation of States in Hilbert Spaces

The function space $L^2(\mathbb{R})$ is the set of all complex-valued functions on the real numbers $\mathbb{R}$ for which

$$f \cdot g = \int_{-\infty}^{\infty} dt\, f^*(t) g(t) \qquad \forall f, g \in L^2(\mathbb{R}) , \qquad [6.25]$$

defines a scalar product between two functions and the norm $\| f \|$ of each function f is finite:

$$\| f \|^2 = \int_{-\infty}^{\infty} dt\, f^*(t) f(t) < \infty . \qquad [6.26]$$

Normed vector spaces in which the norm is given by a scalar product, like $L^2(\mathbb{R})$ or the Euclidean vector space, are called Hilbert spaces (Bronstein et al., 2001). Functions $f \in L^2(\mathbb{R})$ are called square-integrable.

For each Hilbert space $\mathcal{H}$ over a field $\mathbb{C}$, there is a dual space $\mathcal{H}^*$, which is defined as the set of all linear mappings from $\mathcal{H}$ to $\mathbb{C}$. A subset of $L^2(\mathbb{R})^*$ is the set of bilinear forms $\{f_* : L^2(\mathbb{R}) \mapsto \mathbb{C}, f_*(h) = f \cdot h | f \in L^2(\mathbb{R})\}$. This means that for each function f in the Hilbert space, there is a mapping f_* in the dual space, which maps any function h of the Hilbert space to the field $\mathbb{C}$ by taking its scalar product with f. The scalar product $f \cdot g$ on the Hilbert space can therefore be understood not only as an operation on two functions of the Hilbert space, but also as the mapping of g on the field $\mathbb{C}$ via the linear mapping associated with f:

$$f \cdot g = f_*(g) = g_*(f)^* \qquad \forall f, g \in L^2(\mathbb{R}) . \qquad [6.27]$$

Here, the exponential $*$ denotes the complex conjugate.

A subset of a Hilbert space is called a basis if every element of the Hilbert space can be written as a unique linear combination of the elements of the subset. The description of elements of $L^2(\mathbb{R})$ in different bases is used frequently, e.g. when a function is described by its Fourier series: The functions $\left\{ t \mapsto \frac{1}{\sqrt{2\pi}} e^{i\omega t} | \omega \in \mathbb{R} \right\}$ are the basis functions, and the unique coefficients for the linear combination of these basis functions are given by their the scalar product with the respective function $f \in L^2(\mathbb{R})$:

$$\mathcal{F} : L^2(\mathbb{R}) \mapsto L^2(\mathbb{R}) \,, \qquad\qquad [6.28a]$$

$$\mathcal{F}\{f\}(\omega) = \hat{f}(\omega) = \int\limits_{-\infty}^{\infty} dt\, f(t)\, \frac{1}{\sqrt{2\pi}}\, e^{-i\omega t} \,, \qquad\qquad [6.28b]$$

$$\mathcal{F}^{-1}\{\hat{f}\}(t) = f(t) = \int\limits_{-\infty}^{\infty} d\omega\, \hat{f}(\omega)\, \frac{1}{\sqrt{2\pi}}\, e^{i\omega t} \,. \qquad\qquad [6.28c]$$

The function $\hat{f}$ is called the Fourier transform of f.

The Wavelet transform is conceptually similar to the Fourier transform, the only difference being the choice of basis functions: where the Fourier transform uses plane waves, the Wavelet transform uses wavelet functions (see below).

A useful notation for the description of scalar products and changes of bases is the Dirac notation (Cohen-Tannoudji et al., 1977).

Let H be a Hilbert space and an isometric isomorphism (Bronstein et al., 2001) to $L^2(\mathbb{R})$: For each element in $f \in L^2(\mathbb{R})$, define an element $|f\rangle \in H$ in a way that the mapping $L^2(\mathbb{R}) \to H$ is linear and preserves the scalar product:

$$|\alpha f + g\rangle = \alpha|f\rangle + |g\rangle \qquad f,g \in L^2(\mathbb{R}); \, \alpha \in \mathbb{C} \,, \qquad\qquad [6.29a]$$

$$\langle f|g\rangle := |f\rangle \cdot |g\rangle = f \cdot g \qquad f,g \in L^2(\mathbb{R}) \,. \qquad\qquad [6.29b]$$

It follows that

$$\langle f|g\rangle = \langle g|f\rangle^* \quad f,g \in L^2(\mathbb{R}) \,. \qquad\qquad [6.30]$$

The notation $\langle f|g\rangle := |f\rangle \cdot |g\rangle$ implies that $\langle f|$ is a linear operator mapping from H to $\mathbb{C}$ that maps each element $|g\rangle$ of H onto the scalar product of f and g. Therefore, $\langle f|$ is an element of the Dual space H^* of H.

The structures $|\cdot\rangle$ and $\langle\cdot|$ are called 'ket' and 'bra', respectively, which makes the scalar product a 'bra-ket' or bracket. The notation $\langle f|g \rangle$ provides a convenient short form of the scalar product between functions.

Let $\{|\omega\rangle|\omega \in \mathbb{R}\}$ be the subset of H that corresponds to the Fourier basis function subset $\left\{t \mapsto \frac{1}{\sqrt{2\pi}} e^{i\omega t} | \omega \in \mathbb{R}\right\}$ of $L^2(\mathbb{R})$. It is easy to validate that

$$\langle \omega|f \rangle = \hat{f}(\omega) \qquad \forall\, \omega \in \mathbb{R}, f \in L^2(\mathbb{R})\,. \tag{6.31}$$

Therefore, $|f\rangle$ can be interpreted as the function f in general, and $\langle \omega|$ as the linear operator which leads to an evaluation of f at the angular frequency ω. Extending this interpretation, the function value $f(t)$ is just the abstract function f, evaluated at some t (also called the time-domain representation of f):

$$\langle t|f \rangle = f(t)\,, \qquad t \in \mathbb{R}, f \in L^2(\mathbb{R})\,. \tag{6.32}$$

The elements $|t\rangle \in H$ need some corresponding functions in $L^2(\mathbb{R})$, and it can be shown that those 'functions' are δ_t, the Dirac delta distributions centered at t, which in a scalar product with f give the function value $f(t)$.

The Fourier transform can thus be written

$$f(t) = \langle t|f \rangle = \int_{-\infty}^{\infty} d\omega\, \langle t|\omega\rangle\langle\omega|f\rangle = \int_{-\infty}^{\infty} d\omega\, \frac{1}{\sqrt{2\pi}} e^{i\omega t}\, \hat{f}(\omega)\,, \tag{6.33a}$$

$$\hat{f}(\omega) = \langle \omega|f \rangle = \int_{-\infty}^{\infty} dt\, \langle \omega|t\rangle\langle t|f\rangle = \int_{-\infty}^{\infty} dt\, \frac{1}{\sqrt{2\pi}} e^{-i\omega t}\, f(t)\,. \tag{6.33b}$$

The sets of elements $|\omega\rangle$ and $|t\rangle$ are bases of H since they are complete,

$$\int_{-\infty}^{\infty} d\omega\, |\omega\rangle\langle\omega| = \int_{-\infty}^{\infty} dt\, |t\rangle\langle t| = \mathbb{1}_H\,, \tag{6.34}$$

where $\mathbb{1}_H$ is the unit operator on H. Each single operator $|\omega\rangle\langle\omega|$ is a projector on the state $|\omega\rangle$, because the states are normalized:

$$\langle\omega|\omega'\rangle\,d\omega = \delta(\omega-\omega')\,d\omega\,, \qquad [6.35]$$

where δ is the Dirac-distribution (Cohen-Tannoudji et al., 1977).
In this case, the states are also orthogonal, which means that the scalar product between different states is zero. Such bases are called orthonormal bases. However, bases do not have to be orthogonal or normalized, completeness is the only defining condition for a basis.

Wavelet Bases

A wavelet is a function $\varphi \in L^2(\mathbb{R})$ that fulfills the admissibility condition for wavelets (Louis et al., 1998):

$$c_\varphi = 2\pi \int\limits_{-\infty}^{\infty} d\omega\, \frac{|\hat{\varphi}(\omega)|^2}{|\omega|} < \infty\,. \qquad [6.36]$$

A set of basis functions in $L^2(\mathbb{R})$ can be constructed from a single so-called mother wavelet φ by shifting the function on the real axis by an offset $b \in \mathbb{R}$, and by scaling it with a factor $a \in \mathbb{R}\backslash\{0\}$:

$$\left\{ t \mapsto \frac{1}{\sqrt{c_\varphi}}\frac{1}{\sqrt{|a|}}\varphi\left(\frac{t-b}{a}\right) \,|b \in \mathbb{R}, a \in \mathbb{R}\backslash\{0\}\right\}\,. \qquad [6.37]$$

The mappings

$$U(a,b) : L^2(\mathbb{R}) \mapsto L^2(\mathbb{R}^2, d\mu)\,, \qquad [6.38a]$$

$$U(a,b)\{f\}(t) = \frac{1}{\sqrt{|a|}}f\left(\frac{t-b}{a}\right) \quad \forall b \in \mathbb{R}, a \in \mathbb{R}\backslash\{0\}\,, \qquad [6.38b]$$

which induce the shifting and scaling are a representation of the affine group on $L^2(\mathbb{R})$ (Louis et al., 1998), where every group element is defined by a distinct set (a,b). The elements of the Hilbert space H associated with the wavelet basis functions [6.37] are $|a,b\rangle$, given by

$$\langle t|a,b\rangle = \frac{1}{\sqrt{c_\varphi}}U(a,b)\{\varphi\}(t) \,. \tag{6.39}$$

This set of states in H is not necessarily orthogonal, but normalized if φ is a normalized function. It can be shown (Louis et al., 1998) that the states fulfill the necessary property of completeness:

$$\int d\mu(a,b)\,|a,b\rangle\langle a,b| := \int\limits_{-\infty}^{\infty}\frac{da}{a^2}\int\limits_{-\infty}^{\infty}db\,|a,b\rangle\langle a,b| = \mathbb{1}_H \tag{6.40}$$

where $d\mu(a,b) = \frac{da\,db}{a^2}$ is a so-called Haar measure on the space of wavelet transforms $L^2(\mathbb{R}^2, d\mu)$, and $\mathbb{1}_H$ is the identity operator on the Hilbert space.

The wavelet transform can now be defined in analogy to the Fourier transform, Eqs. 6.28:

$$W_\varphi : L^2(\mathbb{R}) \mapsto L^2(\mathbb{R}^2, d\mu) \,, \tag{6.41a}$$

$$(W_\varphi f)(a,b) = \tilde{f}_\varphi(a,b) = \frac{1}{\sqrt{c_\varphi}}\frac{1}{\sqrt{|a|}}\int\limits_{-\infty}^{\infty} dt\,\varphi^*\left(\frac{t-b}{a}\right)f(t) \,, \tag{6.41b}$$

$$(W_\varphi^{-1}\tilde{f}_\varphi)(t) = f(t) = \frac{1}{\sqrt{c_\varphi}}\frac{1}{\sqrt{|a|}}\int d\mu(a,b)\,\varphi\left(\frac{t-b}{a}\right)\tilde{f}_\varphi(a,b) \,. \tag{6.41c}$$

The shorthand Dirac notation makes the derivation of the formulas more clear:

$$\tilde{f}_\varphi(a,b) = \langle a,b|f\rangle = \int\limits_{-\infty}^{\infty} dt\,\langle a,b|t\rangle\langle t|f\rangle \,, \tag{6.42a}$$

$$f(t) = \langle t|f\rangle = \int d\mu(a,b)\,\langle t|a,b\rangle\langle a,b|f\rangle \,. \tag{6.42b}$$

With the help of Fourier basis states of $L^2(\mathbb{R})$, a representation of wavelet coefficients in Fourier space can also be found:

$$\tilde{f}_\varphi(a,b) = \frac{\sqrt{|a|}}{\sqrt{c_\varphi}} \int\limits_{-\infty}^{\infty} d\omega\, e^{i\omega b}\, \hat{\varphi}^*(a\omega)\, \hat{f}(\omega)\,, \qquad [6.43a]$$

$$\hat{f}(\omega) = \frac{1}{\sqrt{c_\varphi}} \int d\mu(a,b)\, \frac{a}{\sqrt{|a|}}\, e^{-i\omega b}\, \hat{\varphi}(\omega a)\, \tilde{f}_\varphi(a,b)\,. \qquad [6.43b]$$

These wavelet transforms are called continuous wavelet transforms (CWT), which means that the group elements of G vary continuously with their parameters (a,b), creating an uncountable set of basis functions.

Properties of Wavelet Transforms

Wavelet functions are well localized in direct and Fourier space (Louis et al., 1998). The localization on the time axis means that only a small portion of the signal f around the wavelet position b is used to compute the wavelet coefficient. Consequently, the positions of certain events in the signal can be detected. This is in contrast to Fourier analysis, where plane waves are used that extend infinitely on the time axis.

The short-time Fourier transform (STFT), also called the windowed Fourier transform, in which the signal is transformed with a windowed part of the plane wave, has a similar advantage of localization. However, the STFT data window is of fixed width, whereas it scales inversely proportional to the frequency in the wavelet transform. This property of the wavelet transform allows for the detection of small-scale structures at high frequencies.

To interpret wavelet transforms, several techniques have been proposed. Using Eq. 6.42b,

$$\int dt\, |f(t)|^2 = \int da\, \tilde{E}_f(a) \qquad [6.44]$$

with

$$\tilde{E}_f(a) = \frac{1}{a^2} \int db\, \left| \tilde{f}_\phi(a,b) \right|^2 . \qquad [6.45]$$

Thereby, when f represents a wind field component, $\tilde{E}_f(a)$ is the distribution of energy per mass over different wavelet scales.

However, shorter wavelets can occur more often than longer ones, therefore a high energy contribution on small scales does not necessarily mean a high contribution for each occurrence. $\tilde{E}_f(a)$ is normalized with a factor $1/a$, which scales with the maximum event occurrence, to arrive at a different energy scale:

$$\tilde{E}_{f,1}(a) = \frac{1}{a} \int db\, \left| \tilde{f}_\varphi(a,b) \right|^2 . \qquad [6.46]$$

The function $\tilde{E}_{f,1}(a)$ gives us the average energy per (possible) event. The function $\tilde{E}_{f,1}(a)$ is often referred to as the scalogram of a CWT (Collineau and Brunet, 1993a).

With these two functions of energy, the most energy-dominant wavelets can be detected: maximizing $\tilde{E}_f(a)$ returns the most important energy scale, maximizing $\tilde{E}_{f,1}(a)$ returns the scale of highest energy per event. In time series analysis, the wavelet transform is usually only evaluated at the dominant energy scale a_0, defined by

$$\tilde{E}_{f,1}(a_0) = \max_a \{ \tilde{E}_{f,1}(a) \} . \qquad [6.47]$$

In this way, only the energetically dominant contributions to the signal are evaluate.

6.2.2. Wavelet-Algorithm for Coherent Structure Detection

To detect a certain signature pattern in a temporal or spatial series, the wavelet analysis should be performed with a mother wavelet which has the general shape of the structure. The wavelet transform will then have a maximum at the position of the overlap. To detect ejections, this analysis follows Segalini and Alfredsson (2012) by using the WAVE wavelet (cf. Fig. 6.14), which is the first derivative of a Gaussian function. The WAVE wavelet has the appropriate shape of an ejection-sweep-cycle in the wind field data, as shown by the ensemble averages of structures measured by Zhang et al. (2011).

The wavelet transform will exhibit a maximum at points where the signal transitions from positive to negative values, thus denoting the 'start' of the ejection. However, instead of using the odd WAVE wavelet, Collineau and Brunet (1993a,b) point out that the detection yields better results using an even MHAT wavelet (Mexican Hat, cf. Fig. 6.14), which exhibits a zero-crossing at the position of the sign change (cf. Fig. 6.12). Therefore, the MHAT is used here to detect structure positions and lengths, but the WAVE wavelet is used to distinguish the 'important' (i.e., those having a large WAVE coefficient) from the 'unimportant' structures.

The MHAT and WAVE mother wavelets are given by

$$\varphi_{\text{WAVE}}(x) = \left(\frac{2}{\pi}\right)^{\frac{1}{4}} (-2x)\, e^{-x^2}\,, \qquad [6.48a]$$

$$\varphi_{\text{MHAT}}(x) = \frac{2}{\sqrt{3}\pi^{\frac{1}{4}}}\, e^{-\frac{1}{2}x^2} \left(1 - x^2\right)\,. \qquad [6.48b]$$

To determine the length scales of the single structures, it has been proposed to use the energetically dominant scales as a measure

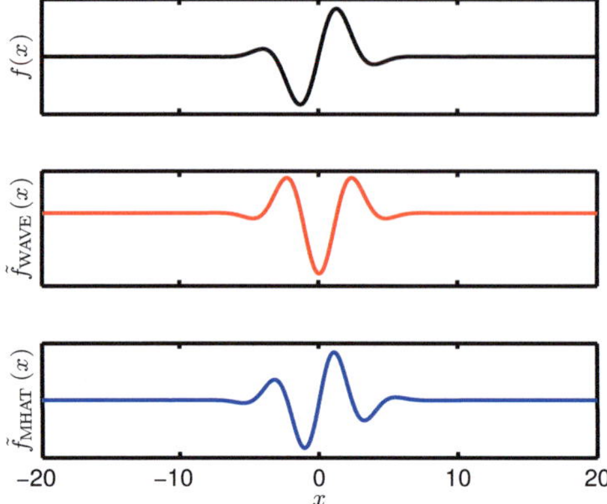

Figure 6.12: Example of a localized event in an exponentially damped sine curve (black) with the wavelet coefficients $\tilde{f}_\varphi(1,x)$ using the WAVE (red) and MHAT (blue) wavelet.

(Collineau and Brunet, 1993a). However, Barthlott et al. (2007) point out that the structure length can best be determined using a ramp length algorithm, which defines the end point of the structure as the point in the time series where the MHAT wavelet transform exhibits the previous maximum. This has the added advantage that the scale on each structure is detected separately, albeit only at the dominant energy scale, instead of only a time series mean.

Wavelet analysis can be applied on spatial data just as well as on time series. Here, the analysis is performed for one-dimensional subsets of the two-dimensional LES and retrieval data, which are oriented in either streamwise or spanwise direction.

For coherent structure detection, therefore the following algorithm is used for the spatial series of wind speed data u and v, which is similar to the algorithms used by Segalini and Alfredsson (2012) and Barthlott et al. (2007):

- Perform the wavelet transforms $\tilde{f}_{\text{MHAT}}$ and $\tilde{f}_{\text{WAVE}}$ of the signal for an appropriate range of scales. To suppress border effects, the signal has to be detrended and zero-padded at both ends (Thomas and Foken, 2005).

- Compute the wavelet spectrum $\tilde{E}_{\text{WAVE},1}(a)$ using Eq. 6.46 and identify the scale of maximum energy per structure, a_0.

- Evaluate $\tilde{f}_{\text{MHAT}}(a_0,b)$ and detect the structure beginnings b_0^i, $i = 1,\ldots,N$, by identifying the zero-crossings with negative slope. This indicates transitions from sweep to ejection.

- To eliminate noise, evaluate $\tilde{f}_{\text{WAVE}}(a_0,b_0^i)$, and reject all detected structures for which $\tilde{f}_{\text{WAVE}}(a_0,b_0) < K \cdot \max_b \tilde{f}_{\text{WAVE}}(a_0,b)$. K has to be pre-defined, with $0 \leq K \leq 1$

- For all valid structure positions, compute the ramp length $L(b_0^i)$, i.e. the distance from b_0 to the consecutive MHAT-maximum.

An example of the results for one 5 km LES streamwise spatial series is shown in Fig. 6.13.

To evaluate time series data, the algorithm must be adapted for the inverted direction of the series, i.e. the MHAT-transform has a positive slope around zero crossings, the WAVE-transform $\tilde{f}_{\text{WAVE}}(a_0,b_0)$ has a minimum at detected structures, and the value must not exceed $K \cdot \min_b \tilde{f}_{\text{WAVE}}(a_0,b)$. Furthermore, the ramp length is defined by the preceding MHAT maximum. Lengths in time series analyses are converted to units of length, as usual, by multiplying with the mean wind.

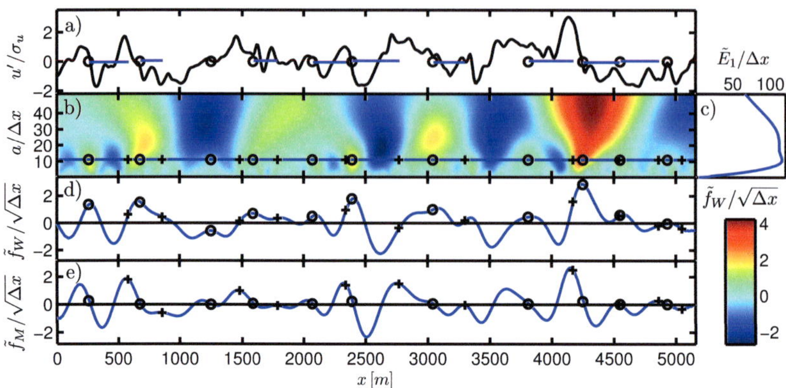

Figure 6.13.: A 1D series $x \mapsto u'(x, y_0, t_0)$ in $u_G = 10$ m/s LES data for some random y_0, t_0, normalized with σ_u (a), its dimensionless WAVE wavelet coefficient $\tilde{f}_W / \sqrt{\Delta x}$ as a function of shift x and scale a(b), the corresponding $\tilde{E}_1$ spectrum (c), the WAVE (d) and MHAT (e) coefficients on the dominant scale. Δx is the distance between adjacent data points in the u' series. Circles denote the beginnings of structures (zero-crossing in MHAT coefficient with negative slope), and +-signs denote structure endings (subsequent maxima in the MHAT coefficients). Normalization with c_φ was neglected for better scaling.

6.2.3. Theoretical Considerations

The influence of the lidar averaging processes on the wavelet coefficients can be assessed using the 1D averaging model of Eq. 6.15 on the spatial series data used for the wavelet transform. The lidar wavelet coefficients for the field f become

$$\tilde{f}_\varphi(a, b) = \frac{1}{\sqrt{c_\varphi}} \frac{1}{\sqrt{|a|}} \int\limits_{-\infty}^{\infty} dx \, \varphi^* \left(\frac{x - b}{a} \right) f_{\mathsf{RET}}(x) \qquad [6.49a]$$

$$= \frac{1}{\sqrt{c_\varphi}} \frac{1}{\sqrt{|a|}} \int\limits_{-\infty}^{\infty} dx \int\limits_{-\infty}^{\infty} dx' \, \varphi^* \left(\frac{x - b}{a} \right) w_{x_0}(x' - x) f_{\mathsf{LES}}(x') \, . \qquad [6.49b]$$

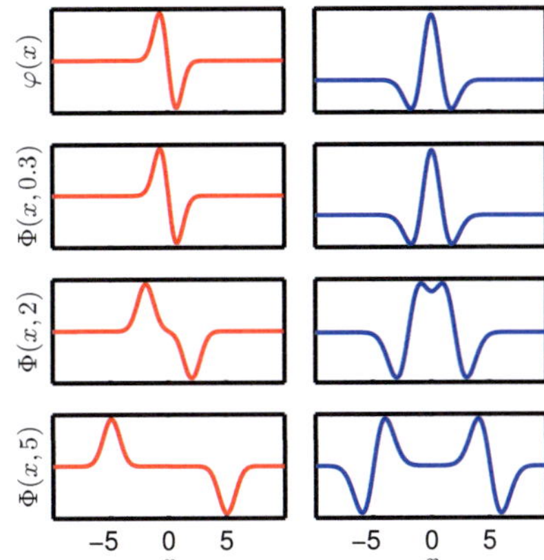

Figure 6.14: The WAVE and MHAT wavelets (top panels, WAVE: red, MHAT: blue), and the resulting effective wavelets (Eq. 6.52) for $a = 1$ and $x_0/a = \{0.3, 2, 5\}$ (top to bottom).

Since x and x' are independent variables, the x-integral can be executed first. Thereby, the effect of the averaging function w_{x_0} is shifted from the field f to the wavelet φ, which results in an effective wavelet Φ:

$$\tilde{f}_\varphi(a,b) = \frac{1}{\sqrt{c_\varphi}} \frac{1}{\sqrt{|a|}} \int_{-\infty}^{\infty} dx' \, \Phi^* \left(\frac{x'-b}{a}, \frac{x_0}{a} \right) f_{\text{LES}}(x') \qquad [6.50]$$

with

$$\Phi \left(\frac{x'-b}{a}, \frac{x_0}{a} \right) = \int_{-\infty}^{\infty} dx \; \varphi \left(\frac{x-b}{a} \right) w_{x_0}(x'-x) . \qquad [6.51]$$

The effect of the averaging process can now be studied by comparing the φ and Φ wavelets. In the case of WAVE and MHAT wavelets φ, Φ can be computed analytically:

$$\Phi_{\text{WAVE}} \left(\frac{x-b}{a}, \frac{x_0}{a} \right) = \frac{a}{2x_0} \left(\frac{2}{\pi} \right)^{\frac{1}{4}} \left[e^{-z^2} \right]_{z=\frac{x-b}{a}-\frac{x_0}{a}}^{z=\frac{x-b}{a}+\frac{x_0}{a}} , \qquad [6.52a]$$

$$\Phi_{\text{MHAT}} \left(\frac{x-b}{a}, \frac{x_0}{a} \right) = \frac{a}{2x_0} \frac{2}{\sqrt{3}\pi^{\frac{1}{4}}} \left[z e^{-\frac{z^2}{2}} \right]_{z=\frac{x-b}{a}-\frac{x_0}{a}}^{z=\frac{x-b}{a}+\frac{x_0}{a}} . \qquad [6.52b]$$

The effective wavelets are functions of $(x-b)/a$ and x_0/a, so the averaging scale x_0 only becomes relevant in relation to the scale a. Φ_{WAVE} and Φ_{MHAT} are shown in Fig. 6.14 for different x_0/a: The effective wavelets remain similar to the original wavelets for small x_0/a, but as soon as the averaging scale becomes larger than the wavelet scale, the wavelet begins to split along their central axis. Those split parts move further apart the larger x_0/a becomes (cf. Fig. 6.15).

For the wavelet analysis of lidar data, this effect means that as soon as the lidar averaging scales are larger than the detected dominant wavelet scales in the spectrum, the resulting wavelet coefficients no longer contain information: The wind field is evaluated at two separate points, so a large overlap with the WAVE-wavelet no longer means an ejection-sweep cycle, but rather an ejection at one point and a sweep some distance away, whereas the information in between is lost. It can therefore be expected that the lidar ramp lengths will match the LES results for scales larger than the lidar averaging scale $2x_0$, whereas for short scales the results will be unreliable.

The ramp length algorithm described above focuses on detected structures on the energetically dominant wavelet scale a_0. However, the breakdown of the lidar spectrum on the high-frequency end also means that the energy contribution to smaller wavelet scales is damped. Therefore, the dominant scale in the retrieval data has an effective lower limit, which will lead to an overestimation for ramp lengths of the LES which fall below the lidar averaging scale.

6.2.4. Results

The analysis was performed as described above for a set of randomly selected spatial series data in x- and y-direction of all u and v wind fields.

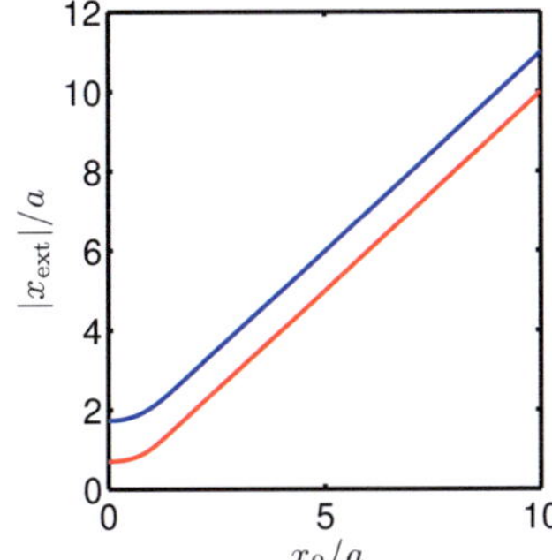

Figure 6.15: Position of the outermost peaks $|x_{ext}|$ in $\Phi_{\mathrm{WAVE}}(x,a,b,x_0)$ (red) and $\Phi_{\mathrm{MHAT}}(x,a,b,x_0)$ (blue) for $b=0$ as a function of the relation between lidar averaging scale x_0 and wavelet scale a.

For the LES data set, one random position y_1 on the y-axis was chosen at each time step t_1, and the wavelet analysis was performed on the resulting 1D series in mean wind direction x, e.g. $x \mapsto u(t_1,x,y_1)$. The number of data points varied for different y_1 due to the rotation in mean wind direction, therefore a restriction was imposed which required the series to have at least 80% of the largest possible number of data points. The series in cross-wind direction were selected in the same way.

In the averaged LES and the retrieval data set, five random series in each direction were choses for wavelet analysis.

Furthermore, the algorithm was applied to time series data from virtual towers in the LES, positioned every 250 m. The length scale results of the time series were converted to spatial scales by multiplying with the mean wind speed.

Ramp lengths L can only be computed when a clear maximum, the dominant energy scale a_0, can be identified in the wavelet spectrum $\tilde{E}_1(a)$. Tab. E.2 in App. E.2 gives an overview of the data set size and data loss due to the absence of a maximum a_0. When a valid maximum was detected, the structure detection algorithm was applied for cutoff-levels $K = \{0, 0.2, 0.4, 0.6, 0.8\}$.

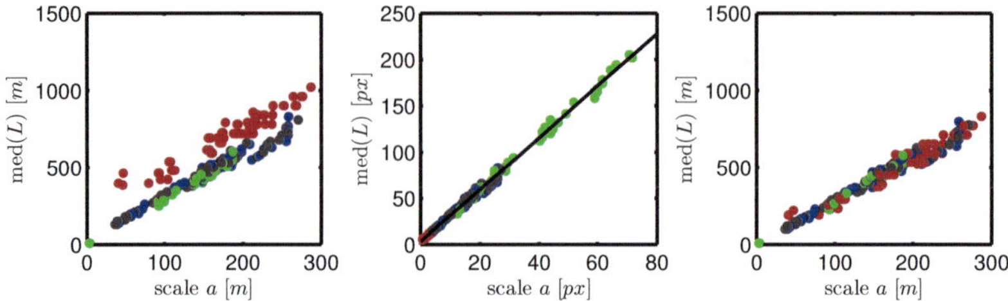

Figure 6.16.: Median ramp lengths in meters calculated from the wavelet algorithm as a function of dominant scales from all data sets (left). The same median ramp lengths in pixels, linear fit: slope $= 2.81$ px, intercept$= 3.13$ px (center). The same ramp length in meters after removing the ordinate-intercept from the pixel-data (right). Colors as in Fig. 6.5.

After the analysis, a rescaling algorithm was applied to the ramp length data. The ramp lengths are defined above as the distances between structure starts b_0 and the following maxima of MHAT. However, this definition allows for a certain inaccuracy, since the position of the maximum is only defined with the accuracy of the data series used. If a bias is introduced here, the length scale over- or underestimation will affect the retrieval data more strongly, since here a 'one-pixel-error' corresponds to scales six or seven times as large as in the LES case. To remove such a bias, the assumption can be made that the mean ramp lengths at scale a are proportional to a,

$$\langle \text{ramp lengths}(a) \rangle \propto a , \qquad [6.53]$$

since a appears as a linear scaling factor in the wavelet functions.
Fig. 6.16 shows the relation between dominant scales a_0 and the mean ramp lengths, which supports the assumption in Eq. 6.53. A linear fit,

$$\langle \text{ramp lengths}(a_0) \rangle = m \cdot a_0 + b \quad \Rightarrow m = 2.81, b = 3.13 , \qquad [6.54]$$

reveals that the length scales in pixel units are over-estimated. The bias of $b = 3.13$ pixels, converted into the metric unit equivalent, is removed from all ramp length data.

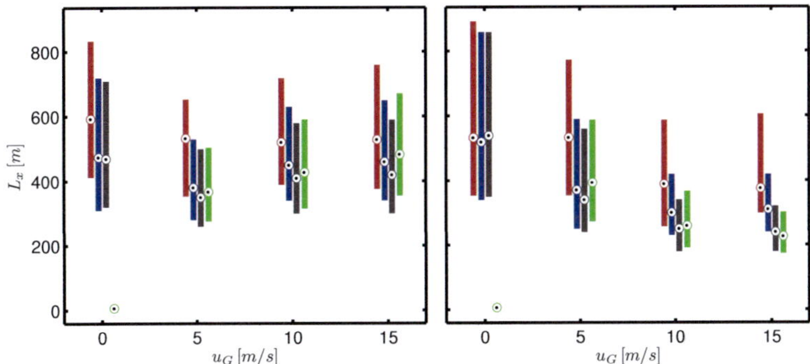

Figure 6.17.: *Wavelet analysis*: Ramp lengths L_x of the u (left) and v (right) wind fields in mean wind direction for all data sets at cutoff $K = 0$. Colors and range as in Fig. 6.5.

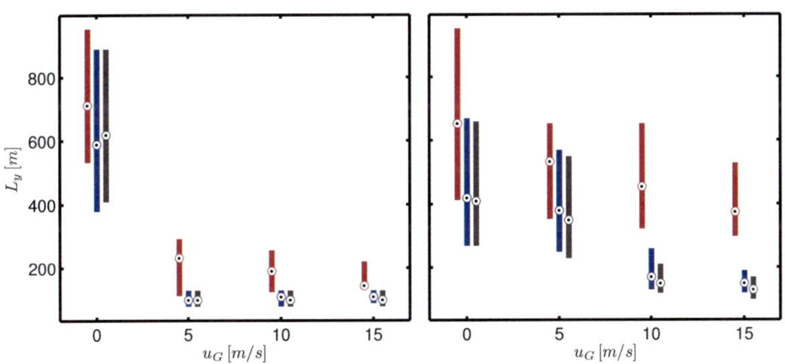

Figure 6.18.: *Wavelet analysis*: Ramp lengths L_y of the u (left) and v (right) wind fields in cross-wind direction for all data sets at cutoff $K = 0$. Colors and range as in Fig. 6.5.

The comparison of LES and retrieval results of length scales is depicted in box plots as in Chap. 6.1, shown in Figs. 6.17 and 6.18. App. G shows the ramp length distributions for the different cutoff levels. Here only the results for $K = 0$ are shown, since the accuracy of the retrieval results is not sensitive to the cutoff value.

The general behavior of the LES wavelet analysis results differs considerably from the integral length scale computations, Figs. 6.5-6.8. Whereas the integral scale of the u fields in x-direction increases with the mean wind, the energetically dominant wavelet scale remains practically constant. The length scale values themselves are higher and have a larger spread. The spread results from the fact that each structure in the series is analyzed, in contrast to the single fixed mean length scale value resulting from the integral length scale algorithm for each time step. The structures are only recorded for the energetically dominant wavelet scales a_0, and the low-pass filtering property of wavelets puts a lower bound on the resulting ramp lengths, which therefore have higher values than the integral scales.

In contrast to the correlation length algorithm, Chap. 6.1, and the clustering algorithm, Chap. 6.3, the L_x and L_y scales are not obtained in pairs for each time step or structure, as in the former and latter case, respectively. Therefore, the anisotropy cannot be measured, but must rather be estimated. Fig. 6.19 shows the range between $p_{25}(L_x)/p_{75}(L_y)$ and $p_{75}(L_x)/p_{25}(L_y)$, where p_n denotes the nth percentile. The centers mark the quotient of median values.

As in the former analysis, the difference between the averaged and fully-resolved LES ramp lengths is negligibly small compared to the data

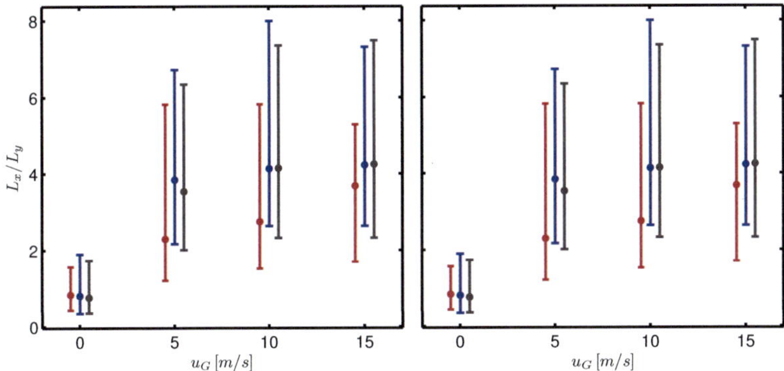

Figure 6.19.: *Wavelet analysis*: Anisotropy L_x/L_y of the u (left) and v (right) wind fields for all data sets at cutoff $K = 0$. Colors as in Fig. 6.5. The bars range from $p_{25}(L_x)/p_{75}(L_y)$ to $p_{75}(L_x)/p_{25}(L_y)$, the centers mark the relation of median values.

spread. The time series data show similarly good results - their high resolution facilitates neither wavelet overestimation nor splitting effects, and the sole analysis of local ramp length neglects possible problems with the ergodic condition.

However, the retrieval results match the LES only for the largest scales, and clearly overestimate for smaller scales, leading again to an underestimation of anisotropy. Furthermore, the length scale spread in the retrieval data becomes larger in comparison for smaller LES ramp lengths.

Both observations agree with the theoretical considerations above: the spread can be explained with the unreliable results due to the wavelet splitting, and the overestimation can be explained by the damped spectrum in high-frequency regions for the LES data. No explanation can be given here why the splitting leads to a strong overestimation.

Fig.6.20 shows the overestimation factor $L_{\mathrm{RET}}/L_{\mathrm{LES}}$ for all data sets of u and v for the different background winds, directions of analysis and cutoff levels. The length scales agree very well for the largest scales. A

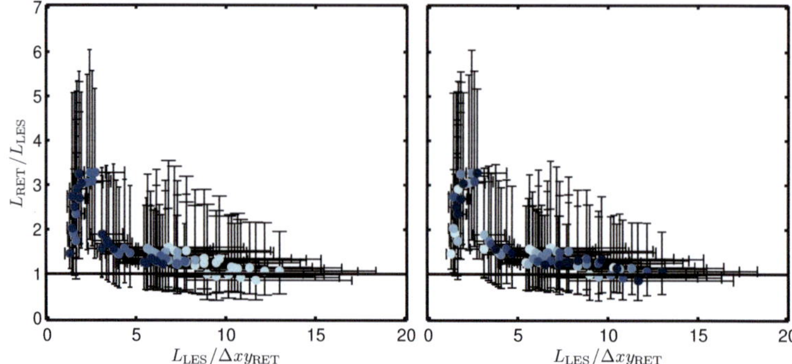

Figure 6.20.: *Wavelet analysis*: Overestimation factor $L_{\text{RET}}/L_{\text{LES}}$ as a function of $L_{\text{LES}}/\Delta xy$ for all wind fields at different wind speeds and cutoff levels K. The colors mark wind speed (darker blue means higher u_G) in the left panel and wavelet cut-off levels in the right panel (darker blue means higher absolute cutoff values). Data points and error bars were computed as in Fig. 6.9.

strong overestimation is visible for L_{LES} smaller than approximately four times the lidar grid resolution, which is the effective lidar averaging scale (cf. Fig. 5.8).

The wavelet splitting effect, resulting in more noise, can be expected for $x_0/a_0 > 1$, which means

$$\frac{L}{\Delta xy} > 2.81 \cdot \frac{x_0}{\Delta xy} , \qquad [6.55]$$

where the numeric factor stems from the linear fit between ramp lengths and scales, Eq. 6.54. This means that the spread should decrease only for values $L_{\text{LES}}/\Delta xy$ of 5.5 or larger. Even though the median values approach unity in this region, the error bars still range from about 0.5 to 2, so large statistics are necessary to estimate ramp lengths correctly.

6.3. Clustering of Low Speed Streaks

The clustering algorithm provides the most descriptive method for structure analysis: By aggregating those regions in the wind field which fall below a certain threshold the typical shapes and scales of structures can be analyzed. No smoothing is involved in this method, which makes it particularly objective, but also sensitive to the lidar averaging effects.

6.3.1. The Clustering Algorithm

The clustering algorithm is used to detect all connected areas in the wind fields in which the wind speed is lower than a certain cutoff level K. The lengths in x- and y-direction are measured as the distance between the outermost points in the cluster.

For discrete wind fields $f = \{f_{(1,1)}, f_{(1,2)}, f_{(2,1)}, \dots\}$, each clusters is a set $C \subset f$, and the length scales L_x, L_y and area A are computed using

$$L_x = \left(\max_{f(i,j) \in C}(i) - \min_{f(i,j) \in C}(i) \right) \cdot \Delta xy, \qquad \text{[6.56a]}$$

$$L_y = \left(\max_{f(i,j) \in C}(j) - \min_{f(i,j) \in C}(j) \right) \cdot \Delta xy, \qquad \text{[6.56b]}$$

$$A = \sum_{f(i,j) \in C} (\Delta xy)^2 . \qquad \text{[6.56c]}$$

This method allows to measure the actual shapes and lengths of low-speed structures, using neither a statistical measure like integral length scales, nor any artificial smoothing like the wavelet algorithms. One example for the clustering is shown in Fig. 6.21.

Time series data can also be subjected to clustering. Here, the structure length is defined as the 1D extent of the cluster, suitably multiplied with the Taylor scales to give units of length. However, a cluster area cannot be defined in this way.

In order to determine average cluster shapes, ensemble averages can be computed: All clusters of one wind field for a certain cutoff level K

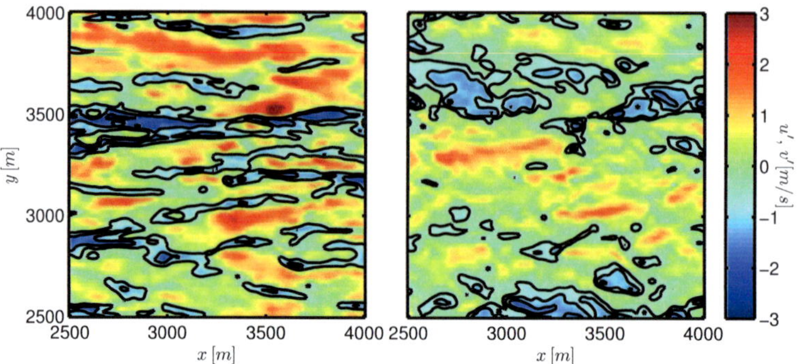

Figure 6.21.: Example of the clustering algorithm: 1.5×1.5 km region in one time step of LES data, fields u (left) and v (right). The contours show clustered ejections with cutoff levels of $\{-2.5, -1.5, -0.5\} \cdot \sigma_u$ and σ_v, respectively.

are normalized with respect to its length and width, so that $L_x = L_y = 1$. Subsequently, the structures are stacked with the centers on top of each other. The map $M(\tilde{x}, \tilde{y})$ is used to count the number of normalized structures which cover a certain point $(\tilde{x}, \tilde{y})$ in the normalized coordinates. A contour line in M at the level $0.5 \cdot \max(M)$ then gives the median shape of the clusters.

6.3.2. Theoretical Considerations

Since the clustering method evaluates the length scales of every single structure without further smoothing or averaging, the effects of the inherent averaging in the retrieval data is expected to be most severe compared to the two other methods of analysis.

Fig. 6.22 shows the effects that are to be expected in the retrieval data due to spatial averaging: The smallest scales in the structures disappear, and larger scales can become even larger. The exact position at

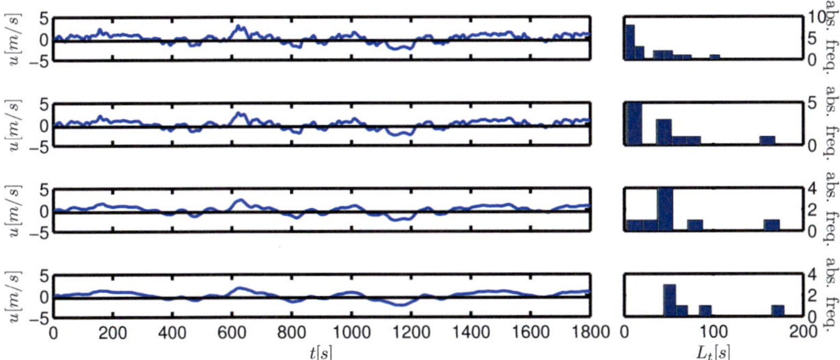

Figure 6.22.: Effect of smoothing on clustering length scales: 30 min time series, smoothed with moving average over $\{1, 10, 30, 60\}$ s (left, top to bottom), and the resulting low-speed cluster lengths below the cutoff-level $K = -0.5 \cdot \sigma_u$.

which any function crosses a certain cutoff level is determined by an interplay of all its spectral components and can therefore not be described by a simple mathematical model.

The length scales L_x measured with virtual towers can be expected to underestimate the LES scales: whereas the spatial analysis measures the maximal extent of a 2D structure, the time series cuts the same structure at a random point y, which will, in most cases, not be the point of largest extent.

6.3.3. Results

The 2D clustering algorithm was applied to all time steps of the u and v data in the averaged LES and the retrieval data sets. Since the method is computationally expensive, only 100 randomly selected time steps were used for the analysis of the LES data. For each selected time step in the respective fields, the wind speed standard deviations (σ_u or σ_v, respectively) were computed, and the clusters aggregated as connected areas

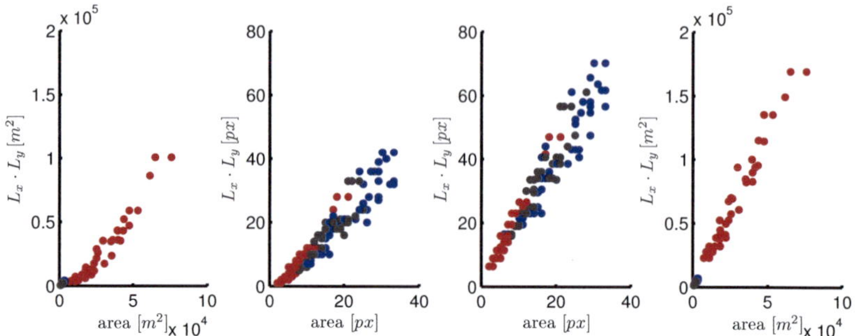

Figure 6.23.: Median cluster length products $L_x \cdot L_y$ in meters from the clustering algorithm as a function of cluster area from all data sets (a), the same data in pixels (b), the same ramp length in pixels after removing the length scale offset -1.52 px from fit (c), the corrected pixel results converted to m^2. Colors as in Fig. 6.5.

in which the wind speed fell below the cutoff level $k \cdot \sigma$. The analysis was performed for $k = \{-3, -2.5, -2, -1.5, -1, -0.5\}$.

The same method was applied to time series data from 500 virtual towers in the LES data, evenly distributed across the area. The large number of towers was possible since 1D-clustering can be executed much quicker than its 2D counterpart.

An overview of the data set size can be found in Tab. E.4. No further criterion had to be fulfilled by the data sets to make the analysis possible, so no data loss occurred.

As in the wavelet case, a rescaling algorithm was applied to the length scales to correct possible pixel errors. Making the assumption that the product of length scales should be proportional to the area of a structure,

$$L_x \cdot L_y \propto A , \tag{6.57}$$

and assuming that the error in both x- and y-direction is given by a constant bias Δ in pixel units, this bias is determined by a least square fit,

$$A = (L_x - \Delta) \cdot (L_y - \Delta) \cdot m , \tag{6.58}$$

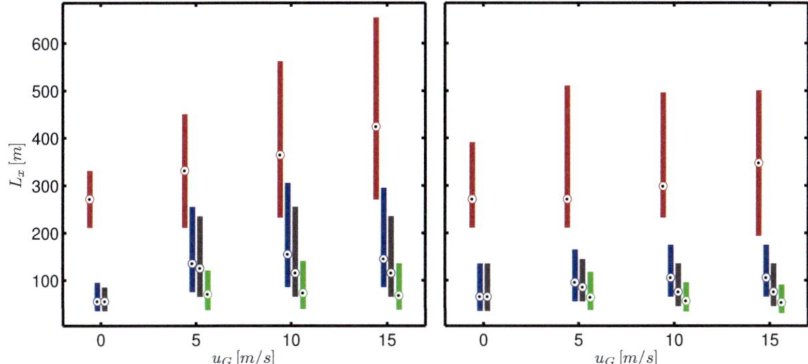

Figure 6.24.: *Clustering*: Structure lengths L_x in mean wind direction of the u (left) and v (right) wind fields for all data sets. Colors and range as in Fig. 6.5.

where the respective L_x, L_y and A are given by the median values for each field for all data sets. The fit yields $\Delta = -1.52$ px, $m = 0.48$ px. With these results, all length scales were corrected in pixel units,

$$L_x \to L_x - \Delta , \tag{6.59a}$$

$$L_y \to L_y - \Delta . \tag{6.59b}$$

and subsequently converted to metric units. The process is shown in Fig. 6.23.

Figs. 6.24 and 6.25 show the length scale distribution in x- and y-direction for a cutoff level -1.5σ. The length scales in the LES data are much smaller than those detected with the correlation length or wavelet algorithm, since here no additional smoothing occurs. Most wind fields exhibit median length scales well below 100 m, which cannot be expected to be resolved by the lidar. The temporal averaging has, again, only a very small overestimation effect which increases with the wind speed. However, the lidar data yield a considerable overestimation.

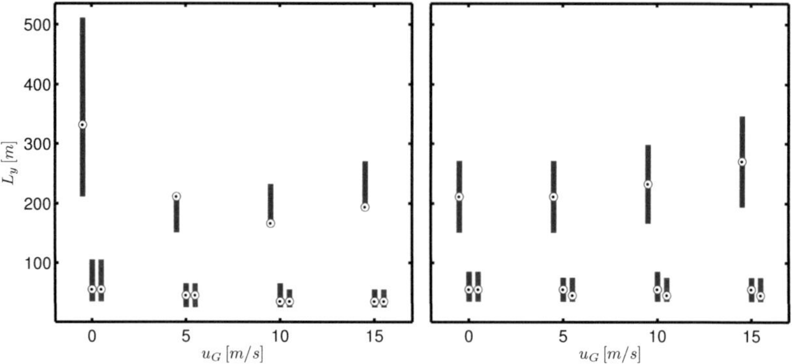

Figure 6.25.: *Clustering*: Structure lengths L_y in cross-wind direction of the u (left) and v (right) wind fields for all data sets. Colors and range as in Fig. 6.5.

The smallest detectable scales appear to be of the order of 150 m. The qualitative development of length scales with the wind speed is hardly detectable in the lidar data.

App. H shows the results for the other cutoff values. It is evident that the quality of results is almost independent of the precise value. The structures become smaller with increasing absolute cutoff, and the data spread decreases due to the fact that for the strictest cutoff values only very few structures were detected (cf. Tab. E.5). Since the overestimation is again most pronounced in the cross-wind direction as here the scales are smallest, the anisotropy is underestimated (Fig. 6.26).

Fig. 6.27 shows a summary of the overestimation factors. Their values as well as their error bars are much larger than the results from any other method, even though the median values almost collapse on a curve. The values become exceedingly large for $L_{LES} < \Delta xy$, which can be expected from the lidar.

Apart from a tendency towards smaller structures for larger cutoff values, the quality of the lidar results does not depend on the cutoff value.

Although the method is capable to determine the precise shape and size of structures, the present scales are much too small to be adequately detected using dual-lidar measurements.

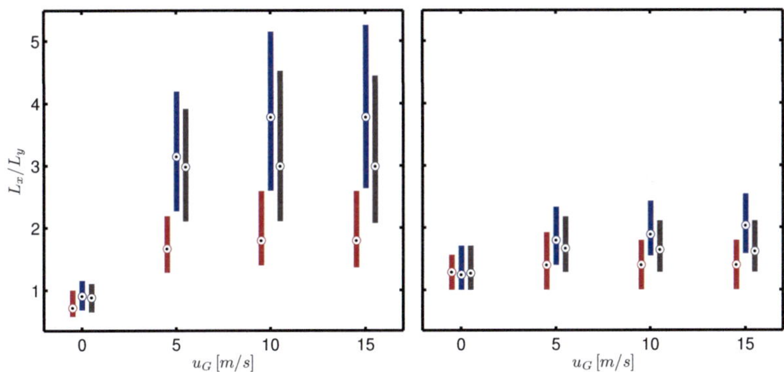

Figure 6.26.: *Clustering*: Structure anisotropy L_x/L_y of the u (left) and v (right) wind fields for all data sets. Colors and range as in Fig. 6.5.

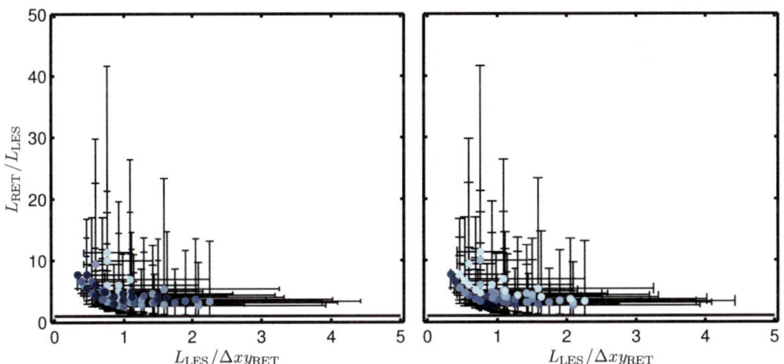

Figure 6.27.: *Clustering*: Overestimation factor L_{RET}/L_{LES} as a function of $L_{LES}/\Delta xy$ from cluster lengths. The colors in the left panel mark wind speed (darker blue means higher u_G) and wavelet cut-off levels in the right panel (darker blue means higher absolute cutoff values). Data points and error bars were computed as in Fig. 6.9.

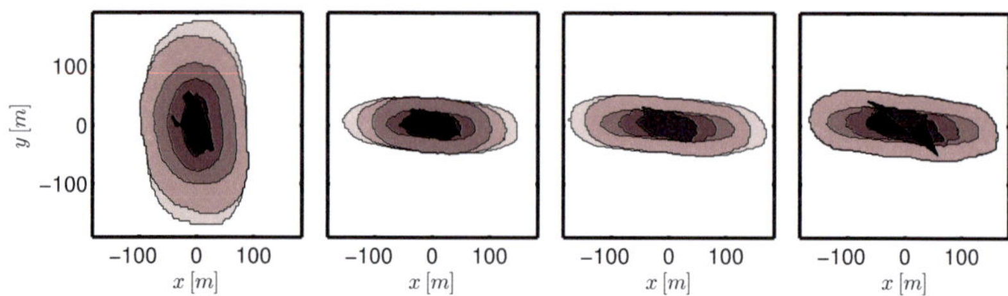

Figure 6.28.: *Clustering:* Overlay of structure shapes (median contours) in retrieval u data for the different cutoff levels k, with u_G increasing from left to right panel. The shapes are stretched to the median length scales for the respective set. Structures from larger absolute cutoff values are shown in darker colors.

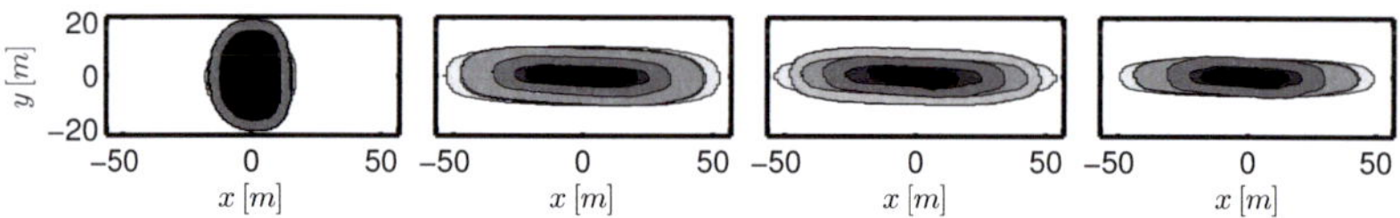

Figure 6.29.: *Clustering:* Overlay of structure shapes in LES u data for the different cutoff levels k, with u_G increasing from left to right panel. Method and colors as in Fig.6.28.

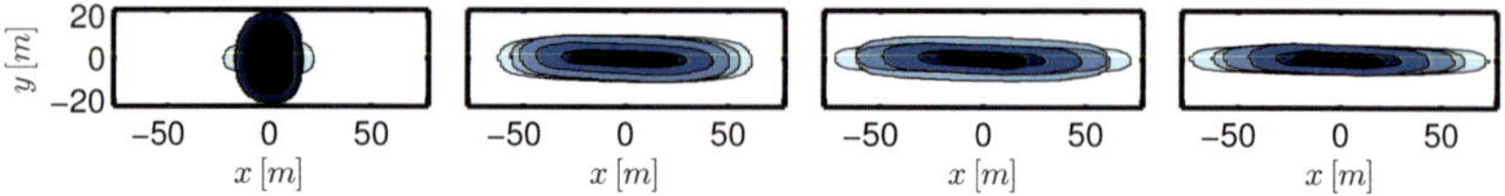

Figure 6.30.: *Clustering:* Overlay of structure shapes in averaged LES u data for the different cutoff levels k, with u_G increasing from left to right panel. Method and colors as in Fig.6.28.

Figs. 6.28-6.30 show the median contours for the different data sets and wind speeds. As expected, the structures are stretched in mean wind direction as the wind speed increases. In the LES and averaged LES data, the median structure scales are largest at an intermediate cutoff value, which is possible when the number of small-scale structures increases rapidly for the cutoff value approaching zero.

Although the length scales are inadequately represented by the lidar results, the general shape and elongation of the structures is qualitatively visible. The shapes at the largest retrieval cutoff values appear uneven due to the very few structures that could be used for ensemble averaging.

6.4. Comparative Results

The spatial averaging in dual-Doppler lidar data necessarily leads to an overestimation in all methods of analysis. Fig. 6.31 shows the comparative overestimation factors. The length scale and resulting anisotropy estimations from dual-lidar data vary considerably in their performance.

The best agreement between LES and virtual lidar results is shown by the correlation length algorithm, especially considering that it is possible to derive the real integral length scales from the lidar data when the spatial averaging processes in the lidar are known, or when further measurements provide a reliable method to measure spatial wind field variances. However, its disadvantage lies in its inability to describe single structures, it is only a statistical measure for the complete time step.

The wavelet algorithm can detect single structures, if only in one dimension, which explains its larger data spread. The performance is very accurate for scales larger than 5 to 6 times the retrieval grid spacing, with median overestimation factors below 1.5. It is mathematically impossible to reliably detect structures with smaller scales or to correct the results.

The clustering algorithm is a theoretically valuable approach since is detects structure features without further smoothing or averaging. However, the detected structures are too small to be accurately detected by the lidar. The spatial averaging on a scale of the structure size leads

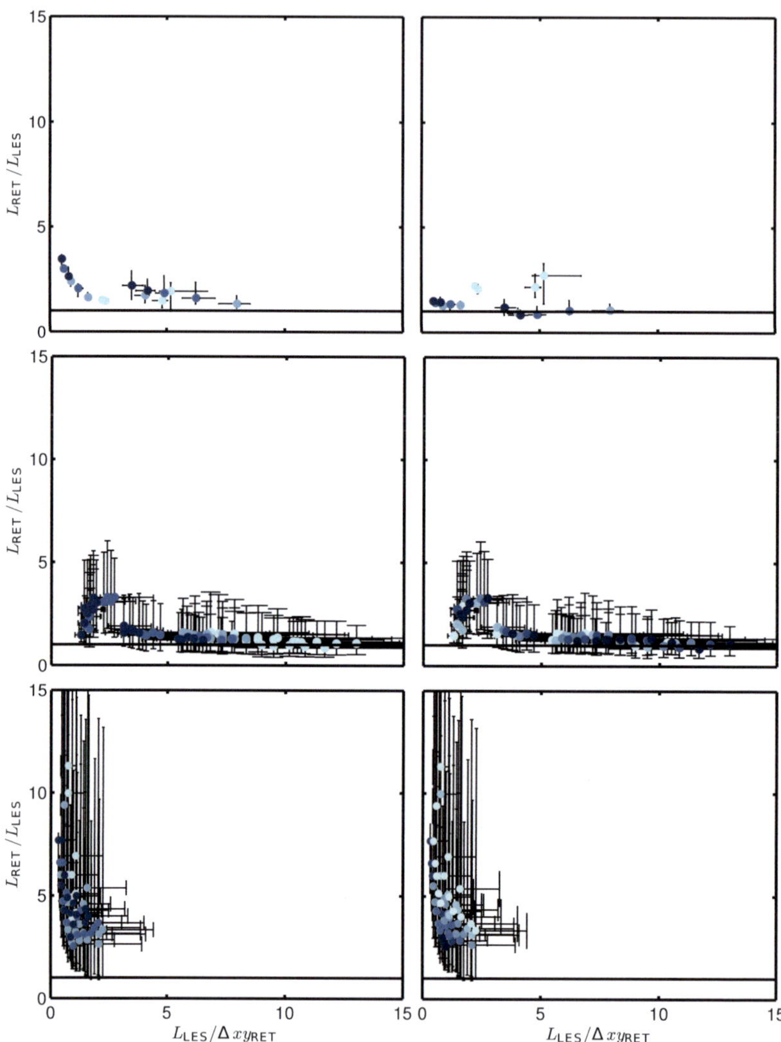

Figure 6.31.: Comparative length scale results:
Integral length scales (top): Overestimation factor L_{RET}/L_{LES} as a function of $L_{LES}/\Delta xy$ from uncorrected (left) and corrected data (right). Method and colors as in Fig. 6.9. *Wavelet analysis* (center): Overestimation factor L_{RET}/L_{LES} as a function of $L_{LES}/\Delta xy$. Method and colors as in Fig. 6.20. *Clustering* (bottom): Overestimation factor L_{RET}/L_{LES} as a function of $L_{LES}/\Delta xy$. Method and colors as in Fig. 6.27. For better visibility, the y-axis was truncated.

to very large overestimation factors and errors. At the moment, no method is available to correct the results.

For the scan times T_0, time averaging has for all methods only a minor contribution to the overestimation.
Time series analyses only yield mean wind direction length scales. The results are generally very accurate, although possible problems due to a lack of ergodicity have to be taken into account for the integral length scales. Time series have the additional disadvantage of providing no spatial data for calm situations, and no information can be obtained about the wind field in spanwise direction.

Both spatial autocorrelation and clustering show a slight tilt of the structures away from the axis of mean wind speed under a small positive angle. Since the LES data used was interpolated to only 10 m height this effect does not have to be significant for surface layers in general. However, it should be noted that the retrieval is able to qualitatively capture this tilt.

7. Techniques for the Derivation of the Vertical Wind Field

*Horizontal lidar scans reproduce only the horizontal wind field compo-
nents. The estimation of vertical momentum fluxes requires further measurements of the vertical wind fields. In this chapter, the Finite Time Lyapunov Exponent is applied to the virtual lidar data, a method which iden-
tifies regions of convergence and divergence in two-dimensional data.
Its applicability for the prediction of near-surface vertical winds is investigated.*

7.1. Finite Time Lyapunov Exponents and Lagrangian Coherent Structures

The following methods and definitions for the Finite Time Lyapunov
Exponent and the Lagrangian Coherent structures were developed by
Shadden et al. (2005), based on earlier works by Haller (2001). A thor-
ough introduction can be found in Shadden (2012).

7.1.1. Lagrangian Coherent Structures

Lagrangian coherent structures (LCS) can be defined for a two-
dimensional vector field (here, the horizontal wind field) over an area

$D \subset \mathbb{R}^2$ and a time interval $[t_0, t_0 + T]$. The wind field determines the trajectory of any particle over the time interval T, always assuming that it moves freely with the wind. Keeping T and t_0 constant, a vector valued mapping ϕ can be defined, where $\phi_{t_0}^{t_0+T}(\mathbf{x})$ is the position of a particle in D at the time $t_0 + T$, which started from $\mathbf{x}$ at t_0:

$$\mathbf{x} \mapsto \phi_{t_0}^{t_0+T}(\mathbf{x}) \, . \tag{7.1}$$

For better readability, the indices t_0 and $t_0 + T$ are omitted in the following.

To obtain a measure for convergence of the vector field, the distance between $\phi(\mathbf{x})$ and the displacement of a particle starting from an infinitesimal distance $\delta\mathbf{x} = \begin{pmatrix} \delta x_1 \\ \delta x_2 \end{pmatrix}$ away from $\mathbf{x}$ is considered:

$$\delta\phi = \phi(\mathbf{x}+\delta\mathbf{x}) - \phi(\mathbf{x}) = \frac{\partial \phi}{\partial x_i}(\mathbf{x}) \, \delta x_i \, . \tag{7.2}$$

Here and hereafter, summation over repeated indices is implied. The absolute value of this displacement is then given by

$$\| \, \delta\phi \, \|^2 = \delta\mathbf{x}^\dagger \, \Delta \, \delta\mathbf{x} \, , \tag{7.3a}$$

$$\text{with } \Delta = \left(\frac{d\phi}{d\mathbf{x}}\right)^\dagger \cdot \left(\frac{d\phi}{d\mathbf{x}}\right) \, , \qquad \text{i.e., } \Delta_{i,j} = \frac{\partial \phi_k^*}{\partial x_i} \frac{\partial \phi_k}{\partial x_j} \, . \tag{7.3b}$$

For constant $\| \, \delta\mathbf{x} \, \|$, the displacement is only a function of the direction of $\delta\mathbf{x}$. Among all possible directions, that one is selected which is aligned with the eigenvector of Δ which has the highest eigenvalue, $\lambda = \max\{\lambda_1, \lambda_2\}$.
Consequently,

$$\| \, \delta\phi \, \|^2 = \lambda \, \| \, \delta\mathbf{x} \, \|^2 \, . \tag{7.4}$$

The *Finite Time Lyapunov Exponent (FTLE)* is defined as (Shadden et al., 2005):

$$\sigma_{t_0}^T(\mathbf{x}) : \ = \frac{1}{|T|} \ln\left(\sqrt{\lambda_{t_0}^{t_0+T}(\mathbf{x})} \right) \, , \tag{7.5}$$

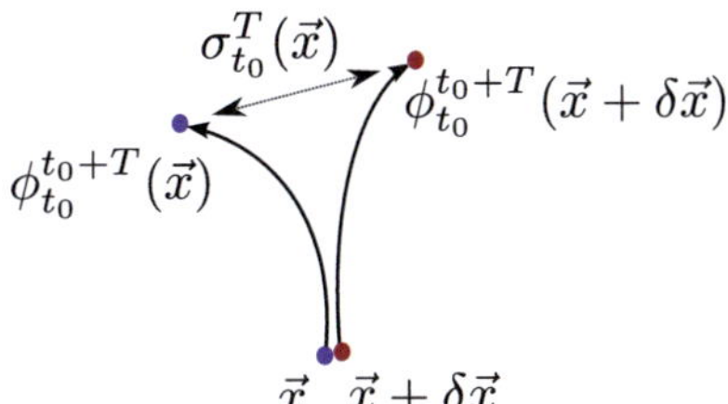

Figure 7.1: Illustration of the FTLE algorithm: Two infinitesimally spaced particles at time t_0 diverge over the time interval T.

where $\lambda_{t_0}^{t_0+T}$ is the aforementioned highest eigenvalue of the Δ matrix, constructed from the $\phi_{t_0}^{t_0+T}$ fields.

Thereby, the FTLE is a measure for convergence or divergence of the trajectories of a vector field. $\sigma < 0$ indicates areas of convergence, while $\sigma > 0$ indicates areas of divergence (cf. Fig. 7.1). If backwards trajectories are computed (i.e., $T < 0$), the meaning of the FTLE-field is reversed, with positive values indicating areas of convergence.

A Lagrangian coherent structure (LCS) is now defined as a ridge in the FTLE-field. A ridge is an injective curve in the 2D plane which fulfills the conditions that (1) the gradient of the FTLE-field is zero perpendicular to the curve, and that (2) the curvature in the same direction is negative, indicating a local maximum.

This means a curve $c : s \mapsto D \subset \mathbb{R}^2, s \in (a,b) \subset \mathbb{R}$, is called a Lagrangian coherent structure of the field $\sigma : D \mapsto \mathbb{R}$ if and only if all of the following conditions apply (Shadden et al., 2005):

- $\forall s \in (a,b) : \nabla\sigma(c(s)) \neq 0 \Rightarrow c'(s) \parallel \nabla\sigma(c(s))$, $\qquad$ [7.6a]

- $\forall s \in (a,b) : \vec{n}(c(s))^T \Sigma(c(s)) \vec{n}(c(s)) < 0$, with $\qquad$ [7.6b]

$$\Sigma : D \mapsto \mathbb{R}^{2\times2}, \ \Sigma_{i,j}(\mathbf{x}) = \frac{\partial^2\sigma}{\partial x_i \partial x_j}(\mathbf{x}) \ \text{the Hessian of } \sigma \text{ and}$$

$\vec{n}(c(s)) \cdot c'(s) = 0$, i.e. $\vec{n}$ points in cross-ridge direction.

Shadden et al. (2005) additionally derived a formula to estimate the flux through the Lagrangian Structures and point out that "for well-defined ridges or ones that rotate at a rate comparable to the local Eulerian field and are computed from a FTLE field which has a sufficiently long integration time, the flux across the LCS is expected to be small". Therefore, LCS can be viewed as barriers in the horizontal flow that shift as the wind field evolves. The lines of convergence denote regions where horizontal air movements are converted to vertical movements, and accordingly lines of divergence denote regions of conversion from vertical to horizontal flow.

7.1.2. The Finite Domain Finite Time Lyapunov Exponent

Tang et al. (2010) point out that the borders of available horizontal wind velocity data regions appear as attractors in the FTLE algorithm, since trajectories stop at the boundaries. To reduce this effect, they proposed a smoothing algorithm, which is also applied here. This algorithm embeds the region of valid data into a larger structure-free background wind field: Let D be the region of valid data, and $D \subset G \subset \mathbb{R}^2$, where the region G is much larger than D. For each time step t, a linear background wind field $\mathbf{v}_L : G \mapsto \mathbb{R}^2$ is defined as the divergence-free wind field which assumes the minimal distance in functional sense from the available wind field $\mathbf{u}$ on the subset D:

$$\mathbf{v}_L(\mathbf{x},t) = \begin{pmatrix} a_{11} & a_{12} \\ a_{21} & a_{11} \end{pmatrix} \cdot \mathbf{x} + \begin{pmatrix} b_1 \\ b2 \end{pmatrix}, \qquad [7.7]$$

where $\mathbf{v}_L$ can be determined by minimizing the functional

$$J[\mathbf{v}] = \sum_{\mathbf{x} \in D} \| \mathbf{v}(\mathbf{x},t) - \mathbf{u}(\mathbf{x},t) \|^2 , \qquad [7.8]$$

hence

$$\min_{\mathbf{v}}\{J[\mathbf{v}]\} = J[\mathbf{v_L}] \, . \qquad\qquad [7.9]$$

The field $\mathbf{v}_L$ is by definition free of LCS.

The smoothing function S is designed to allow the data field $\mathbf{u}$ to fade out into the background linear field around the edges of the data region. Let $dist(\mathbf{x}) : D \mapsto \mathbb{R}$ be the distance of each point in D from the edges of the data region, and Δ the cutoff-distance, then

$$S : G \mapsto \mathbb{R}, S(\mathbf{x}) = \begin{cases} 0, & \mathbf{x} \notin D \\ -2(x/\Delta)^3 + 3(x/\Delta)^2, & \mathbf{x} \in D, dist(\mathbf{x}) \leq \Delta \\ 1, & \mathbf{x} \in D, dist(\mathbf{x}) > \Delta \end{cases} \qquad [7.10]$$

is the lowest order continuously differentiable function that performs the increase from 0 to 1 over the transition region of thickness Δ. The smoothed wind field is then given by

$$\mathbf{u}_{\text{smooth}}(\mathbf{x},t) = \mathbf{v}_L(\mathbf{x},t) + S(\mathbf{x})(\mathbf{u}(\mathbf{x},t) - \mathbf{v}_L(\mathbf{x},t)) \, . \qquad [7.11]$$

The resulting FTLE-fields are also called the 'Finite Domain Finite Time Lyapunov Exponents' (Tang et al., 2010).

7.2. Theoretical Considerations

For measurements close to the ground, it is reasonable to assume a correlation between convergence and upwards movements and between divergence and downwards movements.

A Eulerian approach to estimate the vertical wind can be derived from the incompressibility approximation, $\partial u_i / \partial x_i = 0$, which is exactly true in the LES data (cf. Chap. 3.2.2):

$$w(x,y,z) = -\int_0^z dz' \frac{\partial u}{\partial x}(x,y,z') + \frac{\partial v}{\partial y}(x,y,z') \qquad \text{[7.12a]}$$

$$\approx -z \cdot \left(\frac{\partial u}{\partial x}(x,y,z) + \frac{\partial v}{\partial y}(x,y,z) \right) . \qquad \text{[7.12b]}$$

This finite difference approach becomes less accurate with increasing height z, since it neglects changes in the horizontal divergence in the layers between z and the ground. Furthermore, lidar measurements provide only limited spatial resolutions, so the accuracy of the horizontal divergence results will be limited.

The FTLE, on the other hand, provides a Lagrangian view of wind field convergence. The time integration could improve the estimation of horizontal divergence compared to the finite difference method. However, the lidar data also have a limited time resolution, therefore further errors will be introduced by inaccurate trajectory computations.

A quantitative investigation is needed to determine how accurately the vertical wind field can be deduced from the FTLE field or the horizontal divergence and to determine the relative influences of the advantages and disadvantages of both methods for different integration times T and background wind speeds.

7.3. Results

The FTLE algorithm, as detailed above, was applied to the retrieval data sets, the time averages LES data sets and the full resolution LES data sets. The following steps were executed on the retrieval data sets:

- **Wind field smoothing**

  For each time step in the four retrieval data sets, the smoothed 2D wind field $\mathbf{u}_{\text{RET,smooth}}$ was computed from the retrieved wind $\mathbf{u}_{\text{RET}}$, as described in Chap. 7.1.2, on an area nine times the size of the original retrieval area. The constant Δ, which determines the length scales of the transition between the original wind field in the center and the surrounding linear field $\mathbf{v}_L$ was set to $\Delta = 2 \cdot \Delta xy$. Fig. 7.2 shows the effect of the smoothing around the edges of the data region.

- **Backward trajectories**

  Starting from each time frame t_0 of the retrieval data, 6 trajectories were computed backwards over time intervals of $T = T_0 \cdot \{1,2,3,5,8,13\}$, where T_0 is the time resolution of the retrieval. The end points of the trajectories starting from grid point $\mathbf{x}$, $\phi_{t_0}^{t_0 - T}(\mathbf{x})$ (cf. Eq. 7.1), were determined using finite difference integration of the retrieved wind field:

  With the starting values for position and horizontal wind field,

$$\mathbf{x}_0 = \mathbf{x} , \tag{7.13a}$$

$$\mathbf{u}_0 = \mathbf{u}_{\text{RET,smooth}}(\mathbf{x}_0, t_0) , \tag{7.13b}$$

the stepwise trajectory values are approximated to

$$\mathbf{x}_i = \mathbf{x}_{i-1} - \mathbf{u}_{i-1} \cdot \Delta t \tag{7.14a}$$

$$\mathbf{u}_i = \mathbf{u}_{\text{RET,smooth}}(\mathbf{x}_i, t_0 - i \cdot \Delta t) , \; i = 1, \ldots, T/\Delta t , \tag{7.14b}$$

to obtain the final point

$$\phi_{t_0}^{t_0 - T}(\mathbf{x}) = \mathbf{x}_{T/\Delta t} . \tag{7.15}$$

The integration time step Δt was T_0 for the retrieval data sets, and $\mathbf{u}_{\text{RET,smooth}}(\mathbf{x}_i)$ was determined by linear interpolation of $\mathbf{u}_{\text{RET,smooth}}$ to the point $\mathbf{x}_i$.

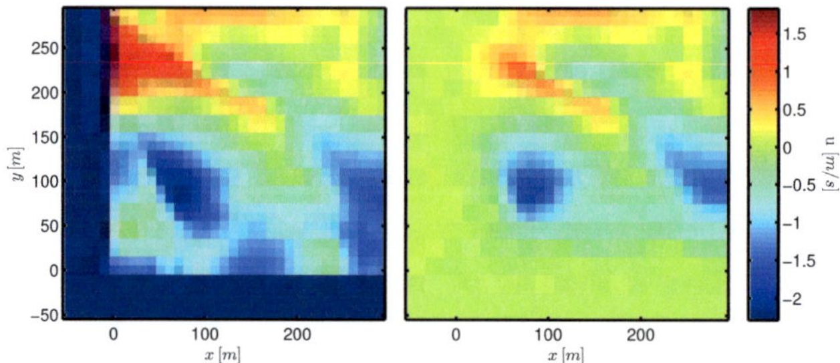

Figure 7.2.: Wind field smoothing for the FTLE algorithm in finite domains, $u_G = 0$ m/s: original (left) and smoothed (right) LES u field at the edge of the data region (no data for $x < 0$ or $y < 0$). The smoothing constant is $\Delta = 2\Delta xy = 120$ m.

- **FTLE**

  From the displacement fields $\phi_{t_0}^{t_0-T}$, the Finite Time Lyapunov Exponent field $\sigma_{t_0}^{T}$ was computed as detailed in Chap. 7.1.1. To this effect, the partial differentials in the Δ-matrix (Eq. 7.3b) were approximated with a finite difference method, i.e.

$$\frac{\partial \phi_k}{\partial x}(n,m) \approx \frac{\phi_k(n+1,m) - \phi_k(n-1,m)}{2\Delta xy} \qquad [7.16a]$$

$$\frac{\partial \phi_k}{\partial y}(n,m) \approx \frac{\phi_k(n,m+1) - \phi_k(n,m-1)}{2\Delta xy} \qquad [7.16b]$$

  where (n,m) are (x,y)-grid-indices and Δxy is the resolution of the wind field grid in meters.

The time averaged LES data were treated in the same way. For the full resolution LES data, the starting times t_0 were chosen as the time steps closest to the retrieval time frames, and the backwards integration was executed with $\Delta t = 1$ s over the same time intervals T, rounded to the full second due to the 1 s LES time resolution.

Furthermore, for each time step in each data set, an approximation for the horizontal divergence was computed. For the LES and time averages LES data, the relative shift of the u, v and w grids (Fig. 3.4) facilitates the computation, using

$$\nabla_H \mathbf{u}_{\text{LES}} \approx \frac{u\left(x+\frac{\Delta x}{2},y\right) - u\left(x-\frac{\Delta x}{2},y\right)}{\Delta x} + \frac{v\left(x,y+\frac{\Delta y}{2}\right) - v\left(x,y-\frac{\Delta y}{2}\right)}{\Delta y} \qquad [7.17]$$

with $\Delta x = \Delta y = \Delta xy_{\text{LES}} = 10$ m.

The retrieval uses the same grid for u and v components, therefore forward differences were used as an approximation:

$$\nabla_H \mathbf{u}_{\text{RET}} \approx \frac{u(x+\Delta x,y) - u(x,y)}{\Delta x} + \frac{v(x,y+\Delta y) - v(x,y)}{\Delta y} . \qquad [7.18]$$

From the horizontal divergence, the approximate vertical wind fields were obtained, using Eq. 7.12:

$$w_{\text{DIVH}}(\mathbf{x}) = -10 \text{ m} \cdot \nabla_H \mathbf{u} . \qquad [7.19]$$

The results for FTLE and horizontal divergence can be compared with the appropriate vertical wind fields. For the LES and time averaged LES, the vertical wind fields chosen for comparison are obviously the full resolution and time averaged LES w fields, respectively. The retrieval data were compared with the time averaged LES w fields in two different ways: firstly, w_{LESAVG} was linearly interpolated on the retrieval axes, and secondly, w_{LESAVG} was smoothed using a moving average over an area of $S_{\text{smooth}} \times S_{\text{smooth}}$ with $S_{\text{smooth}} \approx \Delta xy$ before interpolating it to the retrieval axes. The former case corresponds to point measurements of w (available from towers or vertically staring lidars in measurement data), whereas in the latter case w is averaged down to the lidar scale. The smoothing constant S_{smooth} was chosen as the shortest length spanning an odd number of LES grid cells, i.e. $S_{\text{smooth}} = \{70, 70, 70, 90\}$ m for $u_G = \{0, 5, 10, 15\}$ m/s. Tab. 7.1 summarizes the data sets used for comparison.

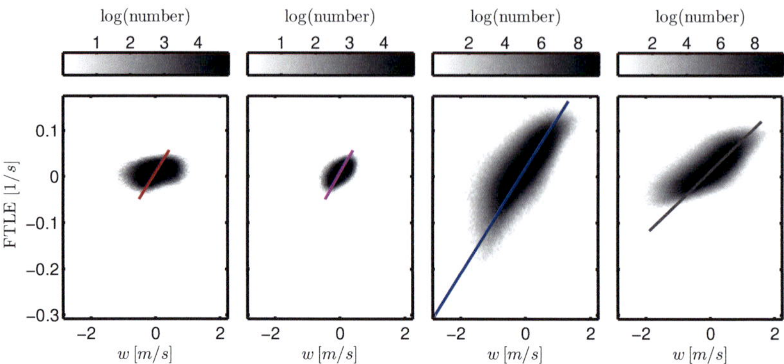

Figure 7.3.: $u_G = 0$ m/s: 2D histograms of vertical winds w and FTLE fields for the data sets I - IV, i. e. the retrieval, the retrieval in comparison with smoothed w, the time averaged LES and full resolution LES data sets (left to right) in logarithmic units in the range $[0, 0.8 \cdot \max\{\text{FTLE}\}]$. The solid lines are the linear fit results.

Since the quantitative relationship between FTLE and the vertical wind is not obvious, the FTLE is converted into an approximate vertical wind field w_{FTLE} by fitting it to the real w field.

Figs. 7.3 and 7.4 show the joint FTLE and w distribution with $u_G = 0\,\text{m/s}$ and $u_G = 10\,\text{m/s}$ for the three data sets. The histograms indicate that the relation between FTLE and w is approximately linear. The colored lines show the results from the linear least squares fit $w = m \cdot \sigma_{t_0}^{t_0 - T} + b$. The fit parameters are summarized in Tabs. I.1 and I.2. The predicted vertical wind from the FTLE is then defined by

$$w_{\text{FTLE}}(\mathbf{x}) = m_{\text{fit}} \cdot \sigma_{t_0}^{t_0 - T} + b_{\text{fit}} \, . \qquad [7.20]$$

#	Data Set	u, v for FTLE computation	w-field	Resolution
I	RET	retrieved wind fields	time avg. LES wind field (intp.)	$\Delta xy, \Delta t = T_0$
II	RET	retrieved wind fields	time avg. LES wind field (intp., sm.)	$\Delta xy, \Delta t = T_0$
III	LESAVG	time avg. LES wind fields	time avg. LES wind field	$\Delta xy_{\text{LES}}, \Delta t = T_0$
IV	LES	LES wind fields	LES wind field	$\Delta xy_{\text{LES}}, \Delta t = 1\,s$

Table 7.1.: Overview of horizontal wind fields used for FTLE computation and the associated vertical wind fields for the three types of data sets.

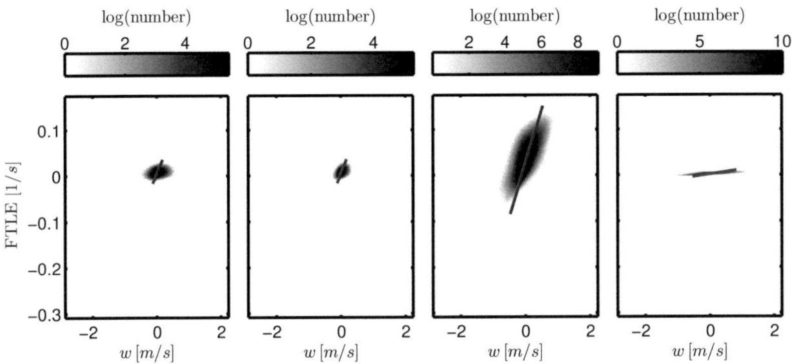

Figure 7.4.: $u_G = 10$ m/s: 2D histograms of vertical winds w and FTLE fields for the data sets I - IV. Method and colors as in Fig. 7.3

Figs. 7.5 and 7.6 show the comparison between the vertical wind, w_DIVH and w_FTLE for an integration time $T = T_0$ for one exemplary time frame in the $u_G = 0$ m/s and the $u_G = 10$ m/s data sets, respectively. While the divergence appears to reproduce w best in the LES and time averaged LES case, the results are less clear in the retrieval case. Both w_DIVH and w_FTLE capture the dominant convergence lines of the smoothed vertical wind field at $u_G = 0$ m/s qualitatively well. Those lines are less apparent in the unsmoothed w-fields, the agreement of which with the predicted wind fields is unsurprisingly less good. For $u_G = 10$ m/s, nearly all structures are averaged out, but w_DIVH appears the reproduce the remaining fluctuations better than w_FTLE.

The correlation coefficients $r_{w,\text{FTLE}}$ and $r_{w,\text{DIVH}}$ between vertical wind fields and the FTLE fields, which were derived from the full data sets, are shown in Fig. 7.8. Furthermore, the mean absolute error (MAE) between the predicted vertical winds and w, normalized with the standard

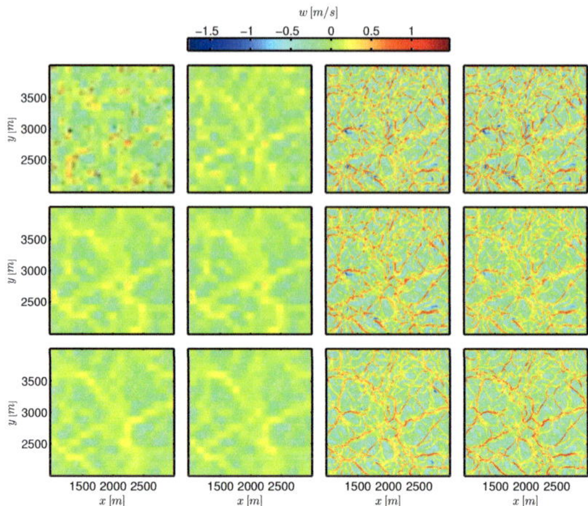

Figure 7.5.: $u_G = 0$ m/s: Comparison between w, w_{DIVH} and w_{FTLE} (top to bottom) for the data sets I - IV, i. e. the retrieval, the retrieval in comparison with smoothed w, time averaged LES and full resolution LES data sets (left to right) on a 2 km×2 km area for one time frame. The FTLE-fields have an integration time of $T = 1 \cdot T_0$.

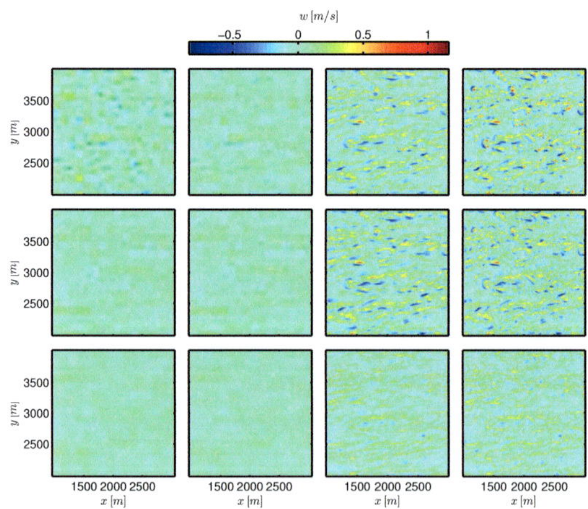

Figure 7.6.: $u_G = 10$ m/s: Comparison between w, w_{DIVH} and w_{FTLE} (top to bottom) for the data sets I - IV, i. e. the retrieval, the retrieval in comparison with smoothed w, time averaged LES and full resolution LES data sets (left to right) on a 2 km×2 km area for one time frame. The FTLE-fields have an integration time of $T = 1 \cdot T_0$.

deviation of w, is shown in Fig. 7.9. The MAE for two fields w_1, w_2 on the same grid with (N_i, N_j, N_k) grid points in the (x, y, t) dimensions is given by

$$\text{MAE}_{w_1,w_2} = \frac{1}{N_i \cdot N_j \cdot N_k} \sum_{i,j,k=1}^{N_i,N_j,N_k} |w_1(i,j,k) - w_2(i,j,k)| . \qquad [7.21]$$

In a direct comparison of both vertical wind prediction methods, the FTLE algorithm outperforms the horizontal divergence in the retrieval case. Therefore one can conclude that the time integration can in part compensate for the low spatial resolution in determining the horizontal divergence. The good agreement between w_{DIVH} and the vertical wind can be seen in the LES results. The deviation can be explained by the necessary rough approximation of the integral, Eq. 7.17. Small errors can also occur due to the interpolation of the u and v fields (Chap. 5.4). In contrast, the LES results show a better performance of the horizontal divergence method compared to the FTLE. It should be noted that $T = T_0$ is the shortest possible integration interval for the retrieval data, whereas the LES data with a time resolution of 1 s could allow for a much shorter integration time.
The LES time averaging appears to have a small positive effect on the result quality.

It is evident that the best agreement of w with w_{FTLE} is always obtained for shorter integration times T.
For all data sets, the errors increases and the correlation decreases for larger u_G in the comparison between w and w_{FTLE}. Note that the theoretical FTLE results (Chap. 7.1.1) are not affected by a constant background wind, rather the rapid change of the wind field appears to introduce larger errors during the trajectory computation. The effects

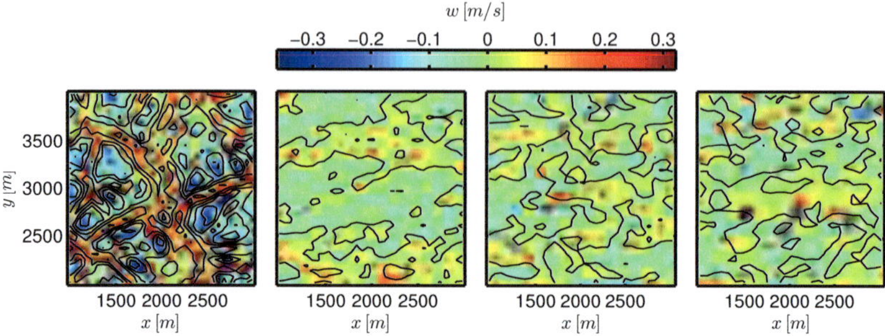

Figure 7.7.: The vertical wind field $w_{LESAVG,intp,sm}$, overlaid with contours of $w_{RET,FTLE}$ at levels {-0.3,-0.2,...,0.2,0.3} m/s.

can be seen in a direct comparison between the smoothed time averages vertical wind field and the $T = T_0$-FTLE-field from the retrieval data in Fig. 7.7: Even though the structures and their shapes are reproduced in the FTLE, their location is increasingly displaced from their position in w. Furthermore, the FTLE fields appear increasingly blurred and unable to resolve small-scale structures.

The effect is not visible in the comparison with w_{DIVH}, which only uses the instantaneous wind field so that the advection has no influence on the results.

Quantitatively, both w_{FTLE} and w_{DIVH} perform rather poorly in predicting the vertical wind from the retrieval data. The absolute error lies between $0.8\,\sigma_w$ and $0.9\,\sigma_w$ for the horizontal divergence when comparing with the vertical point measurements, and even increases for the smoothed vertical wind fields. Additionally, the correlation coefficient is small enough for the fields to be considered uncorrelated. The horizontal divergence method is therefore impractical to predict the vertical wind field.

The FTLE results are more promising, with a correlation coefficient of up to 0.4 for the unsmoothed and up to 0.6 for the smoothed vertical wind

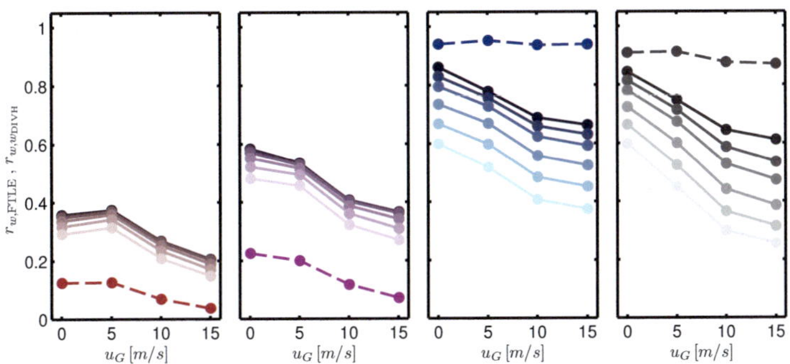

Figure 7.8.: Correlation coefficient between the FTLE and respective vertical wind fields (solid lines), and between w_{DIVH} and the vertical wind fields (dashed lines), for the data sets I - IV, i. e. the retrieval data, the retrieval data in comparison with smoothed w, the time averaged LES and LES data (left to right). The FTLE field results are shown for $\{1,2,3,5,8,13\} \cdot T_0$, where T_0 is the retrieval time constant. Lighter colors mean longer integration times.

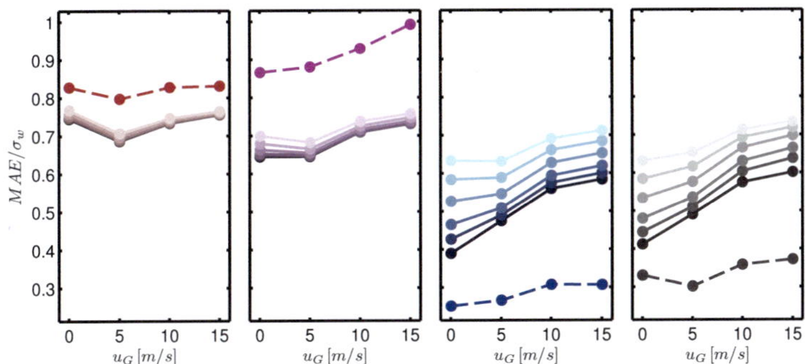

Figure 7.9.: Mean absolute error between the predicted vertical winds w_{FTLE}, w_{DIVH} and the vertical wind w, normalized with the standard deviation of the respective vertical wind field. Colors as in Fig. 7.8.

field. The mean absolute errors remain large, with values between 0.7 and 0.8 σ_w (unsmoothed) and between 0.6 and 0.8 (smoothed), respectively. Nevertheless, the FTLE method proves clearly advantageous for vertical wind field prediction.

Instead of determining the vertical wind speed, the FTLE field can be used to predict the sign rather than the magnitude of w. The further

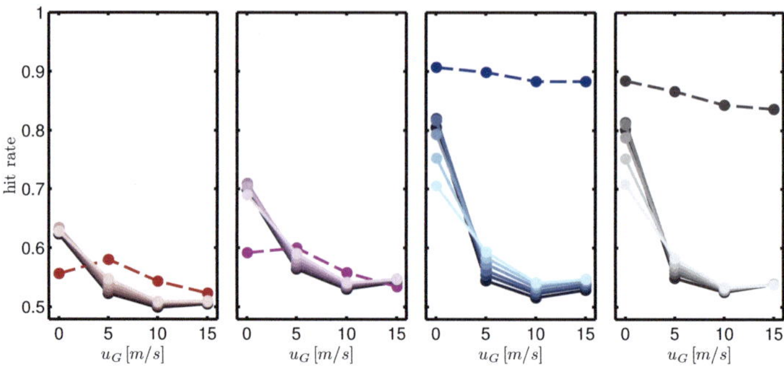

Figure 7.10.: Hit rate for the prediction of sign(w): Number of data points with sign(FTLE) $\equiv$ sign(w) (or sign(w_{DIVH})$\equiv$sign(w)), divided by total number of data points. Colors as in Fig. 7.8.

advantage lies in the theoretical agreement of sign(w) and sign(FTLE) (cf. Eq. 7.5), where no further fit of the FTLE to otherwise obtained vertical wind data is required.

In Fig. 7.10 the performance for this less strict predictand is investigated. The plots show the hit rate, i.e. the number of data points where sign(w) and sign(FTLE) coincide, normalized with the total number of data points (Wilks, 1995). The full contingency tables for the agreement between the signs can be found in Tab. I.3 for the FTLE fields and in Tab. I.4 for the horizontal divergence.

The retrieval results accurately predict the sign of w only in 50% to 65% of all cases (unsmoothed w) and 55% to 70% (smoothed w), respectively. Only for $u_G = 0$ m/s is the hit rate significantly larger than 50%. Interestingly, the FTLE-performance for $u_G > 0$ m/s is not significantly better for the LES and time averaged LES data sets. Even though the visual agreement is better than in the retrieval case (cf. Fig. 7.6), small displacements introduced though time integration errors lowers the hit rate significantly.

7.4. Summary

The Finite Time Lyapunov Exponent is a measure for divergence in two-dimensional wind fields which uses the spread of trajectories using time integration. This Lagrangian approach can be seen in contrast to the Eulerian horizontal divergence computation from a single time step.

The present LES data sets show a good agreement between negative divergence and the vertical wind. However, neither the FTLE nor the horizontal divergence are accurate predictors for the vertical wind, or even its sign, when applied to virtual dual-lidar retrieval data.

The best predictor for w from retrieval is the FTLE-field with the shortest integration time. For the calm situation, it achieves at correlation coefficient of 0.6, a mean absolute error of 0.65 σ_w and a hit rate of 70% when compared to the vertical wind field on the retrieval scale. However, the quality rapidly deteriorates with increasing background wind speed.

Despite the large errors, the FTLE-field and the vertical wind field agree qualitatively well and the Lagrangian coherent structures are clearly visible. The method appears promising for further analyses that do not rely on the exact position but rather the spatial statistics of convergence lines like their shape and intensity distribution or the ridge curvature of the Lagrangian structures. Furthermore, it is conceivable that the structure localization could be improved with enhanced spatial resolution and time integration techniques.

8. Conclusion and Outlook

Coherent structures appear as regular patterns in boundary layer wind fields. They play a significant role in turbulent transport processes in the boundary layer, but their formation and properties are still too little understood to include them in parameterizations of mesoscale atmospheric models.

The applications of Doppler lidars has meant important progress in coherent structure research. With dual-lidar measurements, the two-dimensional wind field can be measured with high time and spatial resolution. However, the measurement results require an independent quality validation.

The goal of this study was to assess the performance of surface layer coherent structure detection techniques on dual-Doppler lidar planar scan data. To this effect, virtual dual-Doppler lidar measurements were performed in four different LES-generated boundary layers.

Virtual lidar measurements were generated from high-resolution LES data. The large-eddy simulations were driven by background geostrophic winds from 0 m/s to 15 m/s and varying surface heat fluxes, and thereby covered a range from purely convective boundary layers without shear to shear-dominated boundary layers.

The visual inspection of the LES data agree with the results from earlier studies: The data show streaks of varying length scales in the sheared surface layer streamwise wind fields, and cellular structures in the vertical wind of the highly convective surface layer.

To perform the virtual measurements, a lidar simulation software tool was developed which operates on given wind fields on a three-dimensional grid. After setting the lidar position, pulse and scan parameters, the software generates radial velocity estimates along the beam from the underlying model data based on the mathematical description of Doppler lidar wind estimation. If the grid constant is small compared to the lidar pulse width, this method yields realistic virtual measurements. Here, the LES had a spatial resolution of 10 m, whereas the pulse width was between 70 m and 90 m.

In dual-Doppler lidar measurements, two lidars scan the same area, and their radial velocity estimates are reassembled to yield two of the three wind components. These dual-lidar measurements were accomplished with the simulator by placing two virtual lidars in the same LES data.

To investigate surface layer structures, a promising approach is to study two-dimensional cross-sections of the wind field. To realize this, the two lidars were programmed to perform synchronized ground-parallel scans on a coplanar area at $z = 10$ m in the LES data sets. The lidar parameters were chosen to match those of the KIT dual-Doppler lidar system. The virtual measurements were preceded by a fundamental error-analysis on dual-Doppler lidar planar scans, which revealed the lidar spatial averaging and time-undersampling as the dominant error sources. Based on this analysis, an optimization scheme for scan patterns was developed which facilitated error-minimization in the scans. The optimized duration T_0 of one planar scan in the given LES set-ups ranged from 11.4 s for high background wind speeds to 14.6 s in the calm situation. In total, each virtual measurement covered a timespan of 30 min and a horizontal area of approximately 15 km^2.

A retrieval algorithm based on a weighted cost function was implemented to convert the virtual dual-lidar data into the horizontal streamwise and spanwise wind components, u and v, on a two-dimensional Cartesian grid in the horizontal lidar scanning plane.

Thereby, the 'real' horizontal wind fields from high-resolution LES and the 'measured' virtual dual-lidar wind fields could be compared directly, including results from wind field based algorithms to classify boundary layer characteristics. This allowed for the first time a quality assessment of dual-Doppler lidar measurements. For further comparison, two other data sets were generated: Firstly, the LES data were time-averaged over the lidar scan intervals to determine the influence of rapidly changing wind fields, and secondly, virtual towers were placed in the LES data to compare the spatial results with high-resolution time series analysis.

Since the dual-lidar data yield only the horizontal wind field, the quantitative analysis was divided into two parts: In the first part the structure length scales in the horizontal wind field were determined, whereas in the second part a derivation of the vertical wind w was attempted.

The first part, Chap. 6, comprises the application of three different structure detection techniques on the horizontal wind fields from lidar, LES, time-averaged LES, and virtual towers. The tower length scales were converted into units of length by multiplication with the mean wind speed.

The first method used the integral over the two-dimensional spatial autocorrelation function to determine the integral length scales L_x and L_y of u and v in streamwise and spanwise direction, respectively. This approach cannot be used to investigate single structures, but rather their mean properties at each time step. Theoretical analyses revealed that the inherent spatial averaging in the dual-lidar data leads to an overestimation of length scales L_{RET} from the lidar compared to the real length scales L_{LES}. Assuming a simplified model for the lidar averaging process, it was found that the length scale overestimation is given by the ratio of variances of the high-resolution and the lidar averaged field,

$$\frac{L_{RET}}{L_{LES}} = \frac{\sigma_{LES}^2}{\sigma_{RET}^2}, \qquad [8.1]$$

which in turn could be expressed as an analytic function of $2x_0/L_{LES}$, where $2x_0$ is the effective lidar spatial averaging scale. The comparison between LES and dual-lidar data agreed with these results and the effective averaging scale was found to be $2x_0 \approx 7\Delta xy$, where Δxy is the cell length of the lidar Cartesian data grid. Further analyses showed that the overestimation factor can be corrected using Eq. 8.1 when σ_{LES}^2 is taken from virtual tower data. An overall value of $L_{RET}/L_{LES} < 1.5$ was achieved, even for L_{LES} as small as $0.5\,\Delta xy$. The correction failed, as expected, for $u_G = 0$ m/s.

This means that integral length scales can be determined accurately for $u_G > 0$ from dual-Doppler lidar measurements, as long as high resolution wind field data from a single meteorological tower are available to perform the scale correction.

The second method used one-dimensional wavelet analysis in both streamwise and spanwise direction on the lidar and LES wind fields, as well as the virtual tower data. Ejection-sweep cycles, i.e., rapid changes from low to high wind speeds, were detected in u with the WAVE and Mexican Hat wavelet. The same algorithm was applied to v. The length scale analysis was only performed on the energetically dominant wavelet scale, where the ramp length was determined for each detected structure. For evaluation, the length scale distributions were compared for the different data sets.

The detected structure lengths L_{RET} in the dual-lidar data again generally overestimate the real lengths, L_{LES}. The results show $L_{\mathrm{RET}}/L_{\mathrm{LES}} \approx 2$ for $L_{\mathrm{LES}} \approx 3\,\Delta xy$, which slowly decreases to one at $L_{\mathrm{LES}} \approx 9\,\Delta xy$. For smaller L_{LES}, the overestimation becomes larger and noisier; it is no longer a monotonous function of $L_{\mathrm{LES}}/\Delta xy$. Theoretical analysis showed that for dominant wavelet scales shorter than the lidar averaging scale the lidar averaging process effectively results in a splitting of the wavelet function into two separate parts located at the borders of the lidar averaging region. With these split functions, a wavelet analysis is no longer meaningful and the results cannot be corrected. On the other hand, wavelet scales much larger than the lidar averaging scales result in a low-pass filtering of also the high-resolution LES data, so that the lidar spatial averaging has no longer any effect on the results.

Consequently, wavelet analysis is a method best suited for the investigation of single structures, as long as an independent measurement is available to ensure that the energetically dominant scales are large enough for the analysis to perform correctly. In all other cases, an interpretation of the wavelet analysis results becomes virtually impossible.

As the third method, a clustering algorithm was applied to the wind components: All coherent regions with u or v smaller than a certain cutoff-value were analyzed in terms of lengths in streamwise and spanwise direction.

The analysis revealed that although the streaky structures appear very elongated in streamwise direction, they are frequently interrupted by small-scale high-speed fluid regions. The clustered low-speed regions are, on average, shorter than $2.5\,\Delta xy$ in any direction. Accordingly, the dual-lidar cannot resolve these small structures, leading to over-estimation factors ranging from 2.5 to 12. Therefore, applying this method of analysis is unsuitable for dual-lidar data under the present circumstances.

The time series data were able to reproduce the streamwise length scale results for all three algorithms for $u_G > 0$. The estimated length scales showed the highest agreement with the LES spatial analysis in the wavelet analysis. The correlation lengths showed a larger error and a negative bias due to reduced ergodicity, and the cluster lengths were underestimated since the tower does not necessarily probe the structures at the point of their longest spanwise extent.

However, time series cannot provide any data in spanwise direction, and the analysis inevitably fails at $u_G = 0$. They should therefore mainly be used to complement spatial analyses.

In the second part, Chap. 7, the Finite Domain Finite Time Lyapunov Exponent (FDFTLE), which is a measure for convergence of the horizontal wind field trajectories, was used to predict the vertical wind under the assumption that updrafts coincide with horizontal wind field convergence close to the ground. The predicted vertical winds w_p showed a high correlation with the vertical winds w from LES in the convective case with $u_G = 0$: $r_{w,w_p} = 0.6$ for the dual-lidar data, and $r_{w,w_p} = 0.8$ for the LES data for a windfield trajectory computation backwards in time over one scan duration T_0. The sign of w was predicted correctly in 70% of all dual-lidar data points, and in 80% of LES data points.

Generally this Lagrangian method showed better results for vertical wind prediction in the lidar data than the simple finite difference integration of the incompressible continuity equation. The prediction quality rapidly decreased with higher u_G and longer trajectory integration times. The FDFTLE is, in summary, a suitable parameter to deduce convective structures in the horizontal wind field data from dual-Doppler lidar, albeit the present time and spatial resolutions result in a more accurate prediction of the sign of w than of the magnitude of w. Therefore, no further structure analyses on the predicted vertical winds were performed.

In all analyses, the time-averaged LES fields performed almost as well as the full-resolution LES. However, the error analysis showed that the error in the radial velocity estimation increases for larger scan times. It is therefore not the duration of the scan, but rather the associated time-undersampling which leads to large errors: In each scan interval of length T_0 the lidar records velocity estimates in each grid cell only for the duration $\Delta t \ll T_0$.

The good agreement between LES and virtual lidar results after the correction of spatial averaging errors suggest that those constitute the dominant contribution to length scale estimation errors from dual-Doppler lidar. The remaining differences may stem from time undersampling or inaccuracies of the simplified spatial averaging model.

In summary, the dual-Doppler lidar planar scan technique is well suited for the investigation of surface layer coherent structures. A dual-Doppler lidar system, in combination with a single meteorological tower, can be used to accurately estimate surface layer integral length scales and the coherent structure statistics on the energetically dominant scale as revealed by wavelet analysis. In calm situations, convective cell structure can be detected with reasonable accuracy. A more precise measurement of w is needed before momentum fluxes $\overline{u'w'}$ can be estimated.

Although the structure detection techniques are not new, their applicability to dual-Doppler lidar data as well as their possible shortcomings and necessary corrections had not yet been investigated. With the results shown here, real dual-Doppler lidar measurements in the atmospheric surface layer can be interpreted reliably.

Several results in this study showed that the main limiting factor for dual-lidar research is the lidar spatial resolution. Unless lasers with shorter wavelength are used, the application of which is limited since they are not eye-safe, the lidar spatial resolution can only be increased at the expense of velocity estimation accuracy or time resolution. The highest achievable spatial resolution at the moment is about 30 m at 1 Hz in Doppler lidars by HALO Photonics, UK (Pearson et al., 2009). This means that Doppler lidars for atmospheric applications will remain at the limit of resolutions required for coherent structures research. The development and general implementation of optimized scans and analyses techniques as shown here is therefore of crucial importance for research results.

As a recipe for coherent structure detection with dual-Doppler lidar measurements, the most important aspects can be summarized as follows:

- The lidar overlap area should cover several square kilometers. Since streaks align in the mean wind direction, the extent in the main wind direction should be at least three kilometers.

- To reduce errors, the lidar intersecting beam angle should be as close as possible to 90° on the lidar overlap area. The lidar elevation should be as small as possible.

- The lidars should scan as fast as possible while maintaining sufficient data density in the outermost scanning area. To achieve this, the optimization method described in Chap. 4.2 can be applied.

- If possible, the lidars should perform synchronized beam sweeps, which doubles the time resolution.

- The lidar measurement should be supplemented with a high-frequency tower wind measurement at lidar measuring height, which can be used for correlation length correction.

- The horizontal wind field retrieval should be performed with a data grid constant Δxy according to the lidar range gate length used for time optimization.

- The correlation length algorithm (Chap. 6.1) can be applied to the resulting horizontal wind fields, giving average streamwise and spanwise correlation lengths for each time frame. The results have to be corrected according to Eq. 6.19. Here, the tower data are necessary to estimate the wind field variance. After correction, the resulting scales can be considered accurate for structures larger than Δxy. The correction fails for calm situations.

- To analyze single structures, the wavelet algorithm (Chap. 6.2) can be applied. The derived length scales are only accurate for structures larger than approximately 5 to 9 times Δxy. For smaller scales, the wavelet analysis fails.

In spring 2013 the first opportunity arose to transfer the theoretical results of this study to real measurements:

The dual-Doppler lidar planar scan technique was implemented during the HOPE experiment (HD(CP)^2 Observational Prototype Experiment)

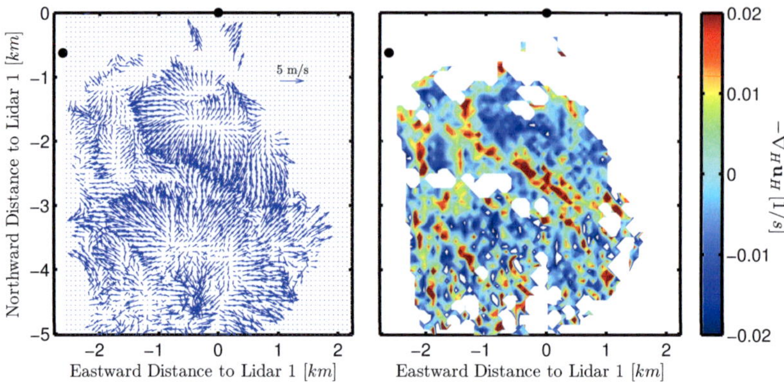

Figure 8.1.: Horizontal wind vectors (left) and convergence $-\nabla_H \mathbf{u}_H$ (right) of the dual-Doppler lidar wind field around 12:07 UTC on April 7, 2013. The black dots denote the lidar positions. The eddy covariance station measured $\overline{w'\theta'} = 0.19$ Km/s, $u_* = 0.3$ m/s, $L_* = -10$ m. The mean wind speed measured by the lidars was $\overline{u}_{\mathrm{RET}} \approx 0.2$ m/s, the lidar detected a boundary layer height of $z_i = 2130$ m.

in Jülich, Germany, as a part of the project "High Definition Clouds and Precipitation for Advancing Climate Prediction". The two KIT Doppler lidars were positioned approximately 2.5 km apart and scanned a coplanar area of about 12 km^2. The scanning plane was elevated 2° due to obstacles, so the mean measurement height was 60 m on average.

As a part of this study, a dual-lidar control software was developed which allowed synchronized beam steering and facilitated the implementation of the optimization algorithm for the scanning patterns. The radial winds were measured over a total period of 300 hours, and the Cartesian horizontal wind components were retrieved using the algorithm developed in this work.

The data analysis, which was not part of this thesis, is still in progress. Figs. 8.1 and 8.2 give an impression of the observed structures: Streaks are visible in the shear-driven surface layer, whereas under convective conditions cell structures of narrow convergence lines occur in the

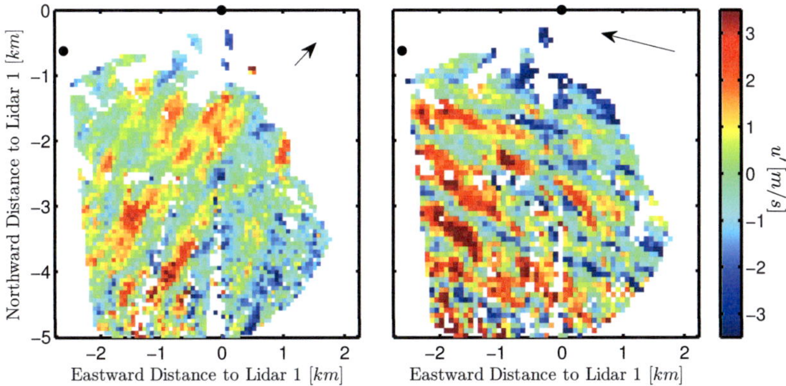

Figure 8.2.: Streamwise wind field component u' of the dual-Doppler lidar wind field around 08:25 UTC on April 17, 2013 (left, $\overline{w'\theta'} = 0.02$ Km/s, $u_* = 0.21$ m/s, $L_* = -27$ m, $\bar{u}_{\mathrm{RET}} = 2.86$ m/s, no measurement for z_i was available), and around 09:05 UTC on April 8, 2013 (right, $\overline{w'\theta'} = 0.14$ Km/s, $u_* = 0.58$ m/s, $L_* = -109$ m, $\bar{u}_{\mathrm{RET}} = 7.45$ m/s, z_i=1000 m). The notation follows Fig. 8.1. The arrows indicate the wind direction in the lidar plane.

horizontal wind field. First results show that, in unstable conditions, the integral length scales derived from the dual-Doppler lidar data vary with the wind speed in the same way as shown in this study. Detailed quantitative analyses of the data will be published shortly, including an integral length scale and wavelet analysis.

The HOPE experiment yielded more than 300 hours of dual-Doppler lidar data. At the same time, several other instruments were deployed: energy balance stations, a 30 m meteorological tower, regularly launched radiosondes, other wind and Raman lidars and a DIAL. The combined data evaluation can be expected to enhance our understanding of surface layer structure formation and characteristics and how these influence and are determined by the state of the boundary layer. Building upon these results the implementation of coherent structure properties in the subgrid-scale parameterization of mesoscale forecast models can be investigated. Since these models usually assume horizontal homogeneity on the subgrid-scale this will require the development of new mathematical and numerical concepts. Considering the important contribution of structures to the surface layer transport their integration in subgrid-models can lead to high advancements for mesoscale modeling.

Since this analysis showed that the horizontal field alone is not sufficient for vertical wind analysis the experimental dual-Doppler lidar results can in the future be extended by the deployment of further wind lidars scanning vertically or in RHI scans to enhance the vertical wind and momentum flux measurements. In combination with measurements from advancing remote-sensing instruments for water-vapor, temperature, and trace gas detection a quantitative characterization and parameterization of surface layer transport processes will soon become possible.

Bibliography

Adrian, R. J., 2007: Hairpin vortex organization in wall turbulence. *Phys. Fluids*, **19 (4)**, 041 301.

Adrian, R. J. and Z. C. Liu, 2002: Observation of vortex packets in direct numerical simulation of fully turbulent channel flow. *J. Visual.*, **5 (1)**, 9–19.

Adrian, R. J., C. D. Meinhart, and C. D. Tomkins, 2000: Vortex organization in the outer region of the turbulent boundary layer. *J. Fluid Mech.*, **422**, 1–54.

Agee, E. M., 1984: Observations from space and thermal convection: A historical perspective. *Bull. Amer. Meteor. Soc.*, **65 (9)**, 938–949.

Alpers, M., R. Eixmann, C. Fricke-Begemann, M. Gerding, and J. Höffner, 2004: Temperature lidar measurements from 1 to 105 km altitude using resonance, Rayleigh, and rotational Raman scattering. *Atmos. Chem. Phys.*, **4 (3)**, 793–800.

Arakawa, A. and V. R. Lamb, 1977: Computational design of the basic dynamical processes of the UCLA general circulation model. *General Circulation Models of the Atmosphere*, Chang, J., Ed., Elsevier, Methods in Computational Physics: Advances in Research and Applications, Vol. 17, 173 – 265.

Baldauf, M., 2008: Stability analysis for linear discretisations of the advection equation with Runge-Kutta time integration. *J. Comput. Phys.*, **227 (13)**, 6638–6659.

Banta, R. M., R. K. Newsom, J. K. Lundquist, Y. L. Pichugina, R. Coulter, and L. Mahrt, 2002: Nocturnal low-level jet characteristics over Kansas during CASES-99. *Bound.-Lay. Metetorol.*, **105 (2)**, 221–252.

Barthlott, C., P. Drobinski, C. Fesquet, T. Dubos, and C. Pietras, 2007: Long-term study of coherent structures in the atmospheric surface layer. *Bound.-Lay. Metetorol.*, **125 (1)**, 1–24.

Barthlott, C., N. Kalthoff, and F. Fiedler, 2003: Influence of high-frequency radiation on turbulence measurements on a 200 m tower. *Meteor. Z.*, **12 (2)**, 67–71.

Breuer, M., 2002: *Direkte numerische Simulation und Large-Eddy-Simulation turbulenter Strömungen auf Hochleistungsrechnern*. Berichte aus der Strömungstechnik, Shaker, Aachen.

Bronstein, I. N., K. A. Semendjajew, G. Musiol, and H. Mühlig, 2001: *Taschenbuch der Mathematik*. 5th ed., Verlag Harri Deutsch, Thun and Frankfurt am Main.

Brooke, J. W. and T. J. Hanratty, 1993: Origin of turbulence-producing eddies in a channel flow. *Phys. Fluids. A - Fluid*, **5 (4)**, 1011–1022.

Browning, K. A. and R. Wexler, 1968: The determination of kinematic properties of a wind field using Doppler radar. *J. Appl. Meteor.*, **7 (1)**, 105–113.

Brümmer, B., 1999: Roll and cell convection in wintertime arctic cold-air outbreaks. *J. Atmos. Sci.*, **56 (15)**, 2613–2636.

Buckingham, E., 1914: On physically similar systems; Illustrations of the use of dimensional equations. *Phys. Rev.*, **4 (4)**, 345–376.

Calhoun, R., R. Heap, M. Princevac, R. Newsom, H. Fernando, and D. Ligon, 2006: Virtual towers using coherent Doppler lidar during the Joint Urban 2003 dispersion experiment. *J. Appl. Meteor.*, **45 (8)**, 1116–1126.

Carlson, J. A., A. Jaffee, and A. J. Wiles, (Eds.) , 2006: *The Millennium Prize Problems*. American Mathematical Society and Clay Mathematics Institute.

Chai, T., C.-L. Lin, and R. K. Newsom, 2004: Retrieval of microscale flow structures from high-resolution Doppler lidar data using an adjoint model. *J. Atmos. Sci.*, **61 (13)**, 1500–1520.

Chakraborty, P., S. Balachandar, and R. J. Adrian, 2005: On the relationships between local vortex identification schemes. *J. Fluid Mech.*, **535**, 189–214.

Cohen-Tannoudji, C., B. Diu, and F. Laloë, 1977: *Quantum mechanics*, Vol. 1. Wiley, New York.

Collier, C. G., F. Davies, K. E. Bozier, A. R. Holt, D. R. Middleton, G. N. Pearson, S. Siemen, D. V. Willetts, G. J. G. Upton, and R. I. Young, 2005: Dual-Doppler lidar measurements for improving dispersion models. *Bull. Amer. Meteor. Soc.*, **86 (6)**, 825–838.

Collineau, S. and Y. Brunet, 1993a: Detection of turbulent coherent motions in a forest canopy part I: Wavelet analysis. *Bound.-Lay. Meteorol.*, **65 (4)**, 357–379.

Collineau, S. and Y. Brunet, 1993b: Detection of turbulent coherent motions in a forest canopy part II: Time-scales and conditional averages. *Bound.-Lay. Metetorol.*, **66 (1-2)**, 49–73.

Davies, F., C. G. Collier, and K. E. Bozier, 2005: Errors associated with dual-Doppler-lidar turbulence measurements. *J. Opt. A: Pure Appl. Op.*, **7 (6)**, S280–S289.

Davies-Jones, R. P., 1979: Dual-Doppler radar coverage area as a function of measurement accuracy and spatial resolution. *J. Appl. Meteor.*, **18 (9)**, 1229–1233.

Deardorff, J. W., 1970: A numerical study of three-dimensional turbulent channel flow at large reynolds numbers. *J. Fluid Mech.*, **41 (2)**, 453–480.

Deardorff, J. W., 1972: Numerical investigation of neutral and unstable planetary boundary layers. *J. Atmos. Sci.*, **29 (1)**, 91–115.

Deardorff, J. W., G. E. Willis, and B. H. Stockton, 1980: Laboratory studies of the entrainment zone of a convectively mixed layer. *J. Fluid Mech.*, **100 (1)**, 41–64.

Demtröder, W., 2009: *Experimentalphysik 3 : Atome, Moleküle und Festkörper.* Springer, Berlin, Heidelberg, New York.

Doms, G., J. Förstner, E. Heise, H.-J. Herzog, D. Mironov, M. Raschendorfer, T. Reinhardt, B. Ritter, R. Schrodin, J.-P. Schulz, and G. Vogel, 2011: A description of the nonhydrostatic regional COSMO model, Part II: Physical parameterization. Tech. rep., Consortium for Small-Scale Modelling.

Drechsel, S., G. J. Mayr, M. Chong, and F. K. Chow, 2010: Volume scanning strategies for 3D wind retrieval from dual-Doppler lidar measurements. *J. Atmos. Ocean. Tech.*, **27 (11)**, 1881–1892.

Drechsel, S., G. J. Mayr, M. Chong, M. Weissmann, A. Dörnbrack, and R. Calhoun, 2009: Three-dimensional wind retrieval: Application of MUSCAT to dual-Doppler lidar. *J. Atmos. Ocean. Tech.*, **26 (3)**, 635–646.

Drobinski, P., R. A. Brown, P. H. Flamant, and J. Pelon, 1998: Evidence of organized large eddies by ground-based Doppler lidar, sonic anemometer and sodar. *Bound.-Lay. Metetorol.*, **88 (3)**, 343–361.

Etling, D., 2008: *Theoretische Meteorologie : Eine Einführung.* 3rd ed., Springer Berlin Heidelberg, Berlin, Heidelberg.

Etling, D. and R. A. Brown, 1993: Roll vortices in the planetary boundary layer: A review. *Bound.-Lay. Metetorol.*, **65 (3)**, 215–248.

Falco, R. E., 1977: Coherent motions in the outer region of turbulent boundary layers. *Phys. Fluids*, **20 (10)**, S124–S132.

Feigenwinter, C. and R. Vogt, 2005: Detection and analysis of coherent structures in urban turbulence. *Theor. Appl. Climatol.*, **81 (3-4)**, 219–230.

Feingold, G., I. Koren, H. Wang, H. Xue, and W. A. Brewer, 2010: Precipitation-generated oscillations in open cellular cloud fields. *Nature*, **466 (7308)**, 849–852.

Foken, T., 2008: The energy balance closure problem: An overview. *Ecol. Appl.*, **18 (6)**, 1351–1367.

Foster, R. C., 1997: Structure and energetics of optimal Ekman layer pertubations. *J. Fluid Mech.*, **333**, 97–123.

Frehlich, R., 1997: Effects of wind turbulence on coherent Doppler lidar performance. *J. Atmos. Ocean. Tech.*, **14 (1)**, 54–75.

Frehlich, R., 2001: Errors for space-based Doppler lidar wind measurements: Definition, performance, and verification. *J. Atmos. Ocean. Tech.*, **18 (11)**, 1749–1772.

Frehlich, R., S. M. Hannon, and S. W. Henderson, 1994: Performance of a 2-μm coherent Doppler lidar for wind measurements. *J. Atmos. Ocean. Tech.*, **11 (6)**, 1517–1528.

Frehlich, R., S. M. Hannon, and S. W. Henderson, 1998: Coherent Doppler lidar measurements of wind field statistics. *Bound.-Lay. Meteorol.*, **86 (2)**, 233–256.

Fröhlich, J., 2006: *Large Eddy Simulation turbulenter Strömungen*. Teubner, Wiebaden.

Grant, H. L., 1958: The large eddies of turbulent motion. *J. Fluid Mech.*, **4 (2)**, 149–190.

Grund, C. J., R. M. Banta, J. L. George, J. N. Howell, M. J. Post, R. A. Richter, and A. M. Weickmann, 2001: High-resolution Doppler lidar for boundary layer and cloud research. *J. Atmos. Ocean. Tech.*, **18 (3)**, 376–393.

Haller, G., 2001: Distinguished material surfaces and coherent structures in three-dimensional fluid flows. *Physica D*, **149 (4)**, 248 – 277.

Hamilton, J. M., J. Kim, and F. Waleffe, 1995: Regeneration mechanisms of near-wall turbulence structures. *J. Fluid Mech.*, **287**, 317–348.

Hartmann, J., C. Kottmeier, and S. Raasch, 1997: Roll vortices and boundary-layer development during a cold air outbreak. *Bound.-Lay. Metetorol.*, **84 (1)**, 45–65.

Heinze, R., 2013: Large-Eddy Simulation von bewölkten Grenzschichten zur Untersuchung von Bilanzen der statistischen Momente zweiter Ordnung und zur Überprüfung von Turbulenzmodellen. Dissertation, Leibniz Universität Hannover, Institute for Meteorology and Climatology.

Hellsten, A. and S. Zilitinkevich, 2013: Role of convective structures and background turbulence in the dry convective boundary layer. *Bound.-Lay. Metetorol.*, **149 (3)**, 323–353.

Henderson, S., P. J. M. Suni, C. Hale, S. Hannon, J. Magee, D. Bruns, and E. Yuen, 1993: Coherent laser radar at 2 μm using solid-state lasers. *IEEE T. Geosci. Remote*, **31 (1)**, 4–15.

Hill, M., R. Calhoun, H. J. S. Fernando, A. Wieser, A. Dörnbrack, M. Weissmann, G. Mayr, and R. Newsom, 2010: Coplanar Doppler lidar retrieval of rotors from T-REX. *J. Atmos. Sci.*, **67 (3)**, 713–729.

Hinze, J. O., 1959: *Turbulence.* 2nd ed., MacGraw-Hill.

Hommema, S. and R. Adrian, 2003: Packet structure of surface eddies in the atmospheric boundary layer. *Bound.-Lay. Metetorol.*, **106 (1)**, 147–170.

Hunt, J. C. R. and J. F. Morrison, 2000: Eddy structure in turbulent boundary layers. *Eur. J. Mech. B - Fluid*, **19 (5)**, 673 – 694.

Hussain, A. K. M. F., 1983: Coherent structures - reality and myth. *Phys. Fluids*, **26 (10)**, 2816–2850.

Inagaki, A. and M. Kanda, 2010: Organized structure of active turbulence over an array of cubes within the logarithmic layer of atmospheric flow. *Bound.-Lay. Metetorol.*, **135 (2)**, 209–228.

Iwai, H., S. Ishii, N. Tsunematsu, K. Mizutani, Y. Murayama, T. Itabe, I. Yamada, N. Matayoshi, D. Matsushima, S. Weiming, T. Yamazaki, and T. Iwasaki, 2008: Dual-Doppler lidar observation of horizontal convective rolls and near-surface streaks. *Geophys. Res. Lett.*, **35 (14)**, L14808.

Jeong, J. and F. Hussain, 1995: On the identification of a vortex. *J. Fluid Mech.*, **285**, 69–94.

Kaimal, J. C. and J. J. Finnigan, 1994: *Atmospheric boundary layer flows : their structure and measurement*. Oxford University Press, New York.

Kaimal, J. C., J. C. Wyngaard, D. A. Haugen, O. R. Coté, Y. Izumi, S. J. Caughey, and C. J. Readings, 1976: Turbulence structure in the convective boundary layer. *J. Atmos. Sci.*, **33 (11)**, 2152–2169.

Kalthoff, N., B. Adler, A. Wieser, M. Kohler, K. Träumner, J. Handwerker, U. Corsmeier, S. Khodayar, D. Lambert, A. Kopmann, N. Kunka, G. Dick, M. Ramatschi, J. Wickert, and C. Kottmeier, 2013: KITcube - a mobile observation platform for convection studies deployed during HYMEX. *Meteor. Z.*, accepted.

Käsler, Y., S. Rahm, R. Simmet, and M. Kühn, 2010: Wake measurements of a multi-MW wind turbine with coherent long-range pulsed Doppler wind lidar. *J. Atmos. Ocean. Tech.*, **27 (9)**, 1529–1532.

Khanna, S. and J. G. Brasseur, 1998: Three-dimensional buoyancy- and shear-induced local structure of the atmospheric boundary layer. *J. Atmos. Sci.*, **55 (5)**, 710–743.

Kim, J. and P. Moin, 1986: The structure of the vorticity field in turbulent channel flow. Part 2. Study of ensemble-averaged fields. *J. Fluid Mech.*, **162**, 339–363.

Kim, S. and S. Park, 2003: Coherent structures near the surface in a strongly sheared convective boundary layer generated by large-eddy simulation. *Bound.-Lay. Metetorol.*, **106 (1)**, 35–60.

Klett, J. D., 1981: Stable analytical inversion solution for processing lidar returns. *Appl. Optics*, **20 (2)**, 211–220.

Kline, S. J., W. C. Reynolds, F. A. Schraub, and P. W. Runstadler, 1967: The structure of turbulent boundary layers. *J. Fluid Mech.*, **30 (4)**, 741–773.

Kottmeier, C., N. Kalthoff, C. Barthlott, U. Corsmeier, J. van Baelen, A. Behrendt, R. Behrendt, A. Blyth, R. Coulter, S. Crewell, P. di Girolamo, M. Dorninger, C. Flamant, T. Foken, M. Hagen, C. Hauck, H. Höller, H. Konow, M. Kunz, H. Mahlke, S. Mobbs, E. Richard, R. Steinacker, T. Weckwerth, A. Wieser, and V. Wulfmeyer, 2008: Mechanisms initiating deep convection over complex terrain during COPS. *Meteor. Z.*, **17 (6)**, 931–948.

Krishnamurthy, R., A. Choukulkar, R. Calhoun, J. Fine, A. Oliver, and K. S. Barr, 2013: Coherent Doppler lidar for wind farm characterization. *Wind Energy*, **16 (2)**, 189–206.

Kristensen, L., P. Kirkegaard, J. Mann, T. Mikkelsen, M. Nielsen, and M. Sjöholm, 2010: Spectral coherence along a lidar-anemometer beam. Tech. Rep. R-1744(EN), Danmarks Tekniske Universitet, Risø Nationallaboratoriet for Bæredygtig Energi.

Kropfli, R. A. and N. M. Kohn, 1978: Persistent horizontal rolls in the urban mixed layer as revealed by dual-Doppler radar. *J. Appl. Meteor.*, **17 (5)**, 669–676.

Lekien, F., C. Coulliette, A. J. Mariano, E. H. Ryan, L. K. Shay, G. Haller, and J. Marsden, 2005: Pollution release tied to invariant manifolds: A case study for the coast of florida. *Physica D*, **210 (1-2)**, 1–20.

Lenschow, D. H., J. Mann, and L. Kristensen, 1994: How long is long enough when measuring fluxes and other turbulence statistics? *J. Atmos. Ocean. Tech.*, **11 (3)**, 661–673.

Lenschow, D. H. and B. B. Stankov, 1986: Length scales in the convective boundary layer. *J. Atmos. Sci.*, **43 (12)**, 1198–1209.

Lenschow, D. H., V. Wulfmeyer, and C. Senff, 2000: Measuring second-through fourth-order moments in noisy data. *J. Atmos. Ocean. Tech.*, **17 (10)**, 1330–1347.

Letzel, M. O., M. Krane, and S. Raasch, 2008: High resolution urban large-eddy simulation studies from street canyon to neighbourhood scale. *Atmos. Environ.*, **42 (38)**, 8770 – 8784.

Lin, C.-L., J. C. McWilliams, C.-H. Moeng, and P. P. Sullivan, 1996: Coherent structures and dynamics in a neutrally stratified planetary boundary layer flow. *Phys. Fluids*, **8 (10)**, 2626–2639.

Lin, C.-L., C.-H. Moeng, P. P. Sullivan, and J. C. McWilliams, 1997: The effect of surface roughness on flow structures in a neutrally stratified planetary boundary layer flow. *Phys. Fluids*, **9 (11)**, 3235–3249.

Lin, C.-L., Q. Xia, and R. Calhoun, 2008: Retrieval of urban boundary layer structures from Doppler lidar data. Part II: Proper orthogonal decomposition. *J. Atmos. Sci.*, **65 (1)**, 21–42.

Lord Rayleigh, 1916: LIX. On convection currents in a horizontal layer of fluid, when the higher temperature is on the under side. *Philosophical Magazine Series 6*, **32 (192)**, 529–546.

Lothon, M., D. H. Lenschow, and S. D. Mayor, 2006: Coherence and scale of vertical velocity in the convective boundary layer from a Doppler lidar. *Bound.-Lay. Metetorol.*, **121 (3)**, 521–536.

Lottman, B. T., R. G. Frehlich, S. M. Hannon, and S. W. Henderson, 2001: Evaluation of vertical winds near and inside a cloud deck using coherent Doppler lidar. *J. Atmos. Ocean. Tech.*, **18 (8)**, 1377–1386.

Louis, A. K., P. Maaß, and A. Rieder, 1998: *Wavelets : Theorie und Anwendungen.* 2nd ed., Teubner-Studienbücher : Mathematik, Teubner, Stuttgart.

Lykossov, V. N. and C. Wamser, 1995: Turbulence intermittency in the atmospheric surface layer over snow-covered sites. *Bound.-Lay. Metetorol.*, **72 (4)**, 393–409.

Mann, J., J.-P. Cariou, M. S. Courtney, R. Parmentier, T. Mikkelsen, R. Wagner, P. Lindelöw, M. Sjöholm, and K. Enevoldsen, 2009: Comparison of 3D turbulence measurements using three staring wind lidars and a sonic anemometer. *Meteor. Z.*, **18 (2)**, 135–140.

Mann, J., A. Peña, F. Bingöl, R. Wagner, and M. S. Courtney, 2010: Lidar scanning of momentum flux in and above the atmospheric surface layer. *J. Atmos. Ocean. Tech.*, **27 (6)**, 959–976.

Maronga, B., 2013: Monin-Obukhov similarity functions for the structure parameters of temperature and humidity in the unstable surface layer: results from high-resolution large-eddy simulations. *J. Atmos. Sci.*, in press.

Mathieu, J. and J. Scott, 2000: *An introduction to turbulent flow.* Cambridge University Press, Cambridge, UK.

Middleton, W. E. K. and A. F. Spilhaus, 1953: *Meteorological Instruments*. 3rd ed., University of Toronto Press, Toronto, 286 S.

Moeng, C.-H. and P. P. Sullivan, 1994: A comparison of shear-and buoyancy-driven planetary boundary layer flows. *J. Atmos. Sci.*, **51 (7)**, 999–1022.

Moin, P. and J. Kim, 1985: The structure of the vorticity field in turbulent channel flow. Part 1. Analysis of instantaneous fields and statistical correlations. *J. Fluid Mech.*, **155**, 441–464.

Newsom, R., R. Calhoun, D. Ligon, and J. Allwine, 2008: Linearly organized turbulence structures observed over a suburban area by dual-Doppler lidar. *Bound.-Lay. Metetorol.*, **127 (1)**, 111–130.

Newsom, R. K., D. Ligon, R. Calhoun, R. Heap, E. Cregan, and M. Princevac, 2005: Retrieval of microscale wind and temperature fields from single- and dual-Doppler lidar data. *J. Appl. Meteor.*, **44 (9)**, 1324–1345.

Pearson, G. N., F. Davies, and C. Collier, 2009: An analysis of the performance of the UFAM pulsed Doppler lidar for observing the boundary layer. *J. Atmos. Ocean. Tech.*, **26 (2)**, 240–250.

Raasch, S., 2014: Palm Group, Institute of Meteorology and Climatology, Leibniz Universität Hannover, http://palm.muk.uni-hannover.de.

Raasch, S. and D. Etling, 1991: Numerical simulation of rotating turbulent thermal convection. *Beitr. Phys. Atmos.*, **64 (3)**, 185–199.

Raasch, S. and D. Etling, 1998: Modeling deep ocean convection: Large eddy simulation in comparison with laboratory experiments. *J. Phys. Oceanogr.*, **28 (9)**, 1786–1802.

Raasch, S. and T. Franke, 2011: Structure and formation of dust devil–like vortices in the atmospheric boundary layer: A high-resolution numerical study. *J. Geophys. Res.*, **116 (D16120)**.

Raasch, S. and M. Schröter, 2001: PALM - a large-eddy simulation model performing on massively parallel computers. *Meteor. Z.*, **10 (5)**, 363–372.

Radlach, M., A. Behrendt, and V. Wulfmeyer, 2008: Scanning rotational Raman lidar at 355 nm for the measurement of tropospheric temperature fields. *Atmos. Chem. Phys.*, **8 (2)**, 159–169.

Riechelmann, T., Y. Noh, and S. Raasch, 2012: A new method for large-eddy simulations of clouds with Lagrangian droplets including the effects of turbulent collision. *New Journal of Physics*, **14 (6)**, 065 008.

Robinson, S. K., 1991: Coherent motions in the turbulent boundary layer. *Annu. Rev. Fluid. Mech.*, **23 (1)**, 601–639.

Röhner, L. and K. Träumner, 2013: Aspects of convective boundary layer turbulence measured by a dual-Doppler lidar system. *J. Atmos. Ocean. Technol.*, **30 (9)**, 2132–2142.

Rothermel, J., C. Kessinger, and D. L. Davis, 1985: Dual-Doppler lidar measurement of winds in the JAWS experiment. *J. Atmos. Ocean. Tech.*, **2**, 138–147.

Schumann, U., 1975: Subgrid scale model for finite difference simulations of turbulent flows in plane channels and annuli. *J. Comput. Phys.*, **18 (4)**, 376–404.

Segalini, A. and P. H. Alfredsson, 2012: Techniques for the eduction of coherent structures from flow measurements in the atmospheric boundary layer. *Bound.-Lay. Metetorol.*, **143 (3)**, 433–450.

Shadden, S., 2012: Lagrangian coherent structures. *Transport and Mixing in Laminar Flows*, Grigoriev, R. and H.-G. Schuster, Eds., Wiley VCH, Weinheim, chap. 3, 59–90.

Shadden, S. C., F. Lekien, and J. E. Marsden, 2005: Definition and properties of Lagrangian coherent structures from finite-time Lyapunov exponents in two-dimensional aperiodic flows. *Physica D*, **212 (3-4)**, 271–304.

Spuler, S. M. and S. D. Mayor, 2005: Scanning eye-safe elastic backscatter lidar at 1.54 μm. *J. Atmos. Ocean. Tech.*, **22 (6)**, 696–703.

Stawiarski, C., K. Träumner, C. Knigge, and R. Calhoun, 2013: Scopes and challenges of dual-Doppler lidar wind measurements - an error analysis. *J. Atmos. Ocean. Tech.*, **30 (9)**, 2044–2062.

Steinfeld, G., 2009: Die Beurteilung von Turbulenzmess- und Analyseverfahren der Mikrometeorologie durch virtuelle Messungen innerhalb von Grobstruktursimulationen. Dissertation, Leibniz Universität Hannover, Institute for Meteorology and Climatology.

Steinfeld, G., S. Raasch, and T. Markkanen, 2008: Footprints in homogeneously and heterogeneously driven boundary layers derived from a Lagrangian stochastic particle model embedded into large-eddy simulation. *Bound.-Lay. Metetorol.*, **129 (2)**, 225–248.

Stull, R. B., 1988: *An Introduction to Boundary Layer Meteorology*, Atmospheric and Oceanographic Sciences Library, Vol. 13, reprint 2009. Springer.

Sykes, R. I. and D. S. Henn, 1989: Large-eddy simulation of turbulent sheared convection. *J. Atmos. Sci.*, **46 (8)**, 1106–1118.

Tang, W., P. W. Chan, and G. Haller, 2010: Accurate extraction of Lagrangian coherent structures over finite domains with application to flight data analysis over Hong Kong International Airport. *Chaos*, **20 (1)**, 017502.

Tang, W., P. W. Chan, and G. Haller, 2011a: Lagrangian coherent structure analysis of terminal winds detected by lidar. Part I: Turbulence structures. *J. Appl. Meteor. Climatol.*, **50 (2)**, 325–338.

Tang, W., P. W. Chan, and G. Haller, 2011b: Lagrangian coherent structure analysis of terminal winds detected by lidar. Part II: Structure evolution and comparison with flight data. *J. Appl. Meteor. Climatol.*, **50 (10)**, 2167–2183.

Thomas, C. and T. Foken, 2005: Detection of long-term coherent exchange over spruce forest using wavelet analysis. *Theor. Appl. Climatol.*, **80 (2-4)**, 91–104.

Thomas, C. and T. Foken, 2007: Flux contribution of coherent structures and its implications for the exchange of energy and matter in a tall spruce canopy. *Bound.-Lay. Metetorol.*, **123 (2)**, 317–337.

Träumner, K., C. Kottmeier, U. Corsmeier, and A. Wieser, 2011: Convective boundary-layer entrainment: Short review and progress using Doppler lidar. *Bound.-Lay. Metetorol.*, **141 (3)**, 369–391.

Wandinger, U., 2005: Introduction to lidar. *Lidar: Range-Resolved Optical Remote Sensing of the Atmosphere*, Weitkamp, C., Ed., Springer, New York, chap. 1, 1–18.

Werner, C., 2005: Doppler wind lidar. *Lidar - Range-Resolved Optical Remote Sensing of the Atmosphere*, Weitkamp, C., Ed., Springer, New York, chap. 12, 325–354.

Wicker, L. J. and W. C. Skamarock, 2002: Time-splitting methods for elastic models using forward time schemes. *Mon. Wea. Rev.*, **130 (8)**, 2088–2097.

Wilks, D. S., 1995: *Statistical methods in the atmospheric sciences*. Academic Press, Inc., San Diego, CA.

Willmarth, W. W. and S. S. Lu, 1972: Structure of the Reynolds stress near the wall. *J. Fluid Mech.*, **55 (1)**, 65–92.

Wulfmeyer, V., A. Behrendt, H.-S. Bauer, C. Kottmeier, U. Corsmeier, A. Blyth, G. Craig, U. Schumann, M. Hagen, S. Crewell, C. Di Girolamo, Paolo andFlamant, M. Miller, A. Montani, S. Mobbs, E. Richard, M. W. Rotach, M. Arpagaus, H. Russchenberg, P. Schlüssel, M. König, V. Gärtner, R. Steinacker, M. Dorninger, D. D. Turner, T. Weckwerth, A. Hense, and C. Simmer, 2008: Research Campaign: the Convective and Orographically induced Precipitation Study: a research and development project of the world weather research program for improving quantitative precipitation forecasting in low-mountain regions. *Bull. Amer. Meteor. Soc.*, **89 (10)**, 1477–1486.

Wulfmeyer, V. and J. Bösenberg, 1998: Ground-based differential absorption lidar for water-vapor profiling: assessment of accuracy, resolution, and meteorological applications. *Appl. Opt.*, **37 (18)**, 3825–3844.

Xia, Q., C.-L. Lin, R. Calhoun, and R. K. Newsom, 2008: Retrieval of urban boundary layer structures from Doppler lidar data. Part I: Accuracy assessment. *J. Atmos. Sci.*, **65 (1)**, 3–20.

Young, G. S., D. A. R. Kristovich, M. R. Hjelmfelt, and R. C. Foster, 2002: Rolls, streets, waves, and more - A review of quasi-two-dimensional structures in the atmospheric boundary layer. *Bull. Amer. Meteor. Soc.*, **83 (7)**, 997–1001.

Zeeman, M. J., W. Eugster, and C. K. Thomas, 2013: Concurrency of coherent structures and conditionally sampled daytime sub-canopy respiration. *Bound.-Lay. Metetorol.*, **146 (1)**, 1–15.

Zhang, Y., H. Liu, T. Foken, Q. L. Williams, M. Mauder, and C. Thomas, 2011: Coherent structures and flux contribution over an inhomogeneously irrigated cotton field. *Theor. Appl. Climatol.*, **103 (1-2)**, 119–131.

Zhou, J., R. J. Adrian, S. Balachandar, and T. M. Kendall, 1999: Mechanisms for generating coherent packets of hairpin vortices in channel flow. *J. Fluid Mech.*, **387**, 353–396.

List of Figures

A. Single Lidar Error Propagation to Dual-Lidar

This appendix chapter is an excerpt from
Stawiarski, Träumner, Knigge, and Calhoun, 2013: *Scopes and Challenges of Dual-Doppler Lidar Wind Measurements - An Error Analysis.* J. Atmos. Ocean. Tech., **30(9)**, 2044-2062. ©*2013 American Meteorological Society. Used with permission.*

A.1. Error Sources in Intersecting Beam Retrieval

The retrieved wind field component in direction of $\hat{\mathbf{e}}_j$ in the lidar plane is given by

$$u_j = \frac{1}{\sin(\Delta(\chi))} \left[rv_1 \sin(\alpha_j + \frac{\Delta\chi}{2}) - rv_2 \sin(\alpha_j - \frac{\Delta\chi}{2}) \right], \qquad \text{[A.1]}$$

where $\Delta\chi$ is the mathematically positive angle measured from $\hat{\mathbf{r}}_1$ to $\hat{\mathbf{r}}_2$ in the plane where $\hat{\mathbf{r}}_1$, $\hat{\mathbf{r}}_2$ are right-handed, and α_j is the detection angle between $\hat{\mathbf{e}}_j$ and $(\hat{\mathbf{r}}_1 + \hat{\mathbf{r}}_2)/2$, likewise measured in the mathematically positive sense.

Single lidar errors occur, if rv_i, az_i and el_i are biased or have a random error. We consider the evaluation plane as fixed, and sum up all radial velocity estimation and angular errors in the variables rv_1, rv_2.

Ideally, the radial velocities are given by

$$rv_1 = \hat{\mathbf{r}}_1 \cdot \mathbf{u} = u_H \cos(\alpha_j - \frac{\Delta\chi}{2} - \gamma_{u_H}), \qquad \text{[A.2a]}$$

$$rv_2 = \hat{\mathbf{r}}_2 \cdot \mathbf{u} = u_H \cos\left(\alpha_j + \frac{\Delta\chi}{2} - \gamma_{u_H}\right), \qquad \text{[A.2b]}$$

where γ_{u_H} is the angle between $\hat{\mathbf{e}}_j$ and the projected wind vector in the plane (again, measured in the positive sense), and u_H is the modulus of said projected wind vector.

The first error source is the velocity estimation itself, leading to a statistical random error on the velocity estimates and a supposedly negligible bias. The second error source, the lidar angles, leads to a shift in lidar beam direction. This shift has a component in the evaluation plane (the in-plane erorr), and a perpendicular component (the out-of-plane error). Both can be propagated to the velocity estimates given by Eqs. A.2.

We assume that the statistical errors and biases in az_i, el_i and rv_i are known and derive their propagation to u_j. This is accomplished by first propagating the angular errors to the rv_i, and subsequently propagating the total rv_i errors to u_j.

A.2. Propagating Angular Errors to rv_i

In contrast to Eqs. A.2, the measured radial velocities are given by

$$rv_1^M = \hat{\mathbf{r}}_1' \cdot \mathbf{u}, \qquad \text{[A.3a]}$$

$$rv_2^M = \hat{\mathbf{r}}_2' \cdot \mathbf{u}, \qquad \text{[A.3b]}$$

where we define $\hat{\mathbf{r}}_1', \hat{\mathbf{r}}_2'$ as the actual lidar beam unit vectors, which deviate from the ideal ones due to angular errors.

To first order in the errors, we have

$$rv_i^M = rv_i + \delta rv_i + \left.\frac{\partial rv_i^M}{\partial az_i}\right|_{\hat{\mathbf{r}}_i' = \hat{\mathbf{r}}_i} \delta az_i + \left.\frac{\partial rv_i^M}{\partial el_i}\right|_{\hat{\mathbf{r}}_i' = \hat{\mathbf{r}}_i} \delta el_i, \qquad \text{[A.4]}$$

with δaz_i, δel_i denoting the angular deviation from the ideal position and δrv_i denoting the lidar random error.

In Eq. A.4, $\hat{\mathbf{r}}'_i$ can be expressed in terms of a right-handed local trihedron $(\hat{\mathbf{r}}_i, \hat{\mathbf{m}}_i, \hat{\mathbf{n}}_n)$: $\hat{\mathbf{r}}_i$ and $\hat{\mathbf{m}}_i$ both lie in the evaluation plane, with $\hat{\mathbf{r}}_i$ being the ideal lidar beam vector withour errors, $\hat{\mathbf{m}}_i$ being perpendicular to $\hat{\mathbf{r}}_i$ and $\hat{\mathbf{n}}_n$ being the plane-normal vector on the evaluation plane (cf. Chap. 4.1.2). For a more formal definition, we use

$$\hat{\mathbf{n}}_n = \frac{\hat{\mathbf{r}}_1 \times \hat{\mathbf{r}}_1}{\| \hat{\mathbf{r}}_1 \times \hat{\mathbf{r}}_2 \|} \, , \tag{A.5a}$$

$$\hat{\mathbf{m}}_i = \hat{\mathbf{n}}_n \times \hat{\mathbf{r}}_i \, . \tag{A.5b}$$

Decomposing $\hat{\mathbf{r}}'_i$ into its parts along the axes of this local orthogonal coordinate system, we find that

$$rv_i^M = \left(\hat{\mathbf{r}}'_i \cdot \hat{\mathbf{r}}_i \right) rv_i + \left(\hat{\mathbf{r}}'_i \cdot \hat{\mathbf{n}}_n \right) \left(\hat{\mathbf{n}}_n \mathbf{u} \right) + \left(\hat{\mathbf{r}}'_i \cdot \hat{\mathbf{m}}_i \right) \left(\hat{\mathbf{m}}_i \mathbf{u} \right) . \tag{A.6}$$

Careful computation shows that

$$\left. \frac{\partial \hat{\mathbf{r}}'_i}{\partial az_i} \right|_{\hat{\mathbf{r}}'_i = \hat{\mathbf{r}}_i} = \hat{\mathbf{r}}_i \times \hat{\mathbf{k}} \, , \tag{A.7a}$$

$$\left. \frac{\partial \hat{\mathbf{r}}'_i}{\partial el_i} \right|_{\hat{\mathbf{r}}'_i = \hat{\mathbf{r}}_i} = \frac{1}{\cos(el_i)} \left(\hat{\mathbf{r}}_i \times \hat{\mathbf{k}} \right) \times \hat{\mathbf{r}}_i = \frac{1}{\cos(el_i)} \left(\hat{\mathbf{k}} - \sin(el_i)\hat{\mathbf{r}}_i \right) \, , \tag{A.7b}$$

and from this we derive the expressions for the scalar products:

$$\left. \frac{\partial \hat{\mathbf{r}}'_i}{\partial az_i} \right|_{\hat{\mathbf{r}}'_i = \hat{\mathbf{r}}_i} \cdot \hat{\mathbf{n}}_n = \hat{\mathbf{m}}_i \cdot \hat{\mathbf{k}} \, , \tag{A.7c}$$

$$\left. \frac{\partial \hat{\mathbf{r}}'_i}{\partial az_i} \right|_{\hat{\mathbf{r}}'_i = \hat{\mathbf{r}}_i} \cdot \hat{\mathbf{m}}_i = -\hat{\mathbf{n}}_n \cdot \hat{\mathbf{k}} \, , \tag{A.7d}$$

$$\left. \frac{\partial \hat{\mathbf{r}}'_i}{\partial el_i} \right|_{\hat{\mathbf{r}}'_i = \hat{\mathbf{r}}_i} \cdot \hat{\mathbf{n}}_n = \frac{1}{\cos(el_i)} \hat{\mathbf{n}}_n \cdot \hat{\mathbf{k}} \, , \tag{A.7e}$$

$$\left. \frac{\partial \hat{\mathbf{r}}'_i}{\partial el_i} \right|_{\hat{\mathbf{r}}'_i = \hat{\mathbf{r}}_i} \cdot \hat{\mathbf{m}}_i = \frac{1}{\cos(el_i)} \hat{\mathbf{m}}_i \cdot \hat{\mathbf{k}} \, . \tag{A.7f}$$

With these results, we find from Eq. A.6:

$$\left.\frac{\partial rv_i^M}{\partial az_i}\right|_{\hat{\mathbf{r}}_i'=\hat{\mathbf{r}}_i} = (\hat{\mathbf{m}}_i \cdot \hat{\mathbf{k}})\,(\hat{\mathbf{n}}_n \cdot \mathbf{u}) - (\hat{\mathbf{n}}_n \cdot \hat{\mathbf{k}})\,(\hat{\mathbf{m}}_i \cdot \mathbf{u}) \,, \qquad \text{[A.8a]}$$

$$\left.\frac{\partial rv_i^M}{\partial el_i}\right|_{\hat{\mathbf{r}}_i'=\hat{\mathbf{r}}_i} = \frac{(\hat{\mathbf{n}}_n \cdot \hat{\mathbf{k}})\,(\hat{\mathbf{n}}_n \cdot \mathbf{u}) + (\hat{\mathbf{m}}_i \cdot \hat{\mathbf{k}})\,(\hat{\mathbf{m}}_i \cdot \mathbf{u})}{\cos(el_i)} \,. \qquad \text{[A.8b]}$$

We can write down explicit expressions for the scalar products:

$$\hat{\mathbf{n}}_n \cdot \mathbf{u} = u_\perp \,, \qquad\qquad\qquad\qquad \text{[A.9a]}$$

$$\hat{\mathbf{m}}_i \cdot \mathbf{u} = -u_H \sin(\alpha_j \mp \frac{\Delta\chi}{2} - \gamma_{u_H}) \,, \qquad \text{[A.9b]}$$

$$\hat{\mathbf{n}}_n \cdot \hat{\mathbf{k}} = \cos(\gamma_z) \,, \qquad\qquad\qquad \text{[A.9c]}$$

$$\hat{\mathbf{m}}_i \cdot \hat{\mathbf{k}} = \pm\frac{1}{|\sin(\Delta\chi)|}\left[\sin(el_{i'}) - \sin(el_i)\cos(\Delta\chi)\right]. \qquad \text{[A.9d]}$$

Here, $u_\perp$ is the wind speed perpendicular to the evaluation plane, γ_z is the angle between the plane-normal vector $\hat{\mathbf{n}}_n$ and the z-axis $\hat{\mathbf{k}}$. The upper sign applies if the respective lidar i is Lidar 1, the lower sign applies if it is Lidar 2. The index i' indicates the other lidar.

It should be noted that the expressions in Eq. A.8 each contain two summands, one of which scales with the perpendicular wind speed $u_\perp$, and the other of which scales with the in-plane wind, u_H. Those two components arise from the angular error contributions to out-of-plane tilt or in-plane tilt, respectively. In the error propagation below, we will cosider both contributions separately.

A.3. Statistical Error Propagation

The random error in the retrieved wind field component is given by

$$
[\sigma_{DD}^{\text{single}}(u_j)]^2 = \sum_{i=1,2} \left(\frac{\partial u_j}{\partial rv_i}\right)^2 (\sigma_i^{rv})^2
$$

$$
= \frac{\sin^2(\alpha_j + \frac{\Delta\chi}{2})}{\sin^2(\Delta\chi)}(\sigma_1^{rv})^2 + \frac{\sin^2(\alpha_j - \frac{\Delta\chi}{2})}{\sin^2(\Delta\chi)}(\sigma_2^{rv})^2 \ . \qquad [\text{A.10}]
$$

The statistical radial velocity variances consists of the random error of velocity estimation, the variance due to in-plane angular errors and the variance due to out-of-plane errors. The latter two are not statistically independent, because both have contributions from elevation and azimuth angles. It is therefore necessary to also consider their covariance. Nevertheless, the splitting is advisable, since the in-plane error is the only part that can be estimated with the measurement results alone, i.e., the wind speed in the lidar plane. Without the splitting, no quantitative statement respective the angular errors can be made at all.

$$
(\sigma_i^{rv})^2 = \left(\sigma_i^{rv,rnd}\right)^2 + \left(\sigma_i^{rv,ip}\right)^2 + \left(\sigma_i^{rv,oop}\right)^2 + \text{cov}(ip,oop) \ . \qquad [\text{A.11}]
$$

The random part is known. The in-plane and out-of-plane parts and their covariance can be traced back to the contributing angles using the results from the previous section:

$$
\left(\sigma_i^{rv,ip}\right)^2 = u_H^2 \sin^2(\alpha_j \mp \frac{\Delta\chi}{2} - \gamma_{u_H}) \qquad\qquad [\text{A.12}]
$$

$$
\cdot \left\{ \cos^2(\gamma_z)\,(\sigma_i^{az})^2 + \frac{(\sin(el_{i'}) - \sin(el_i)\cos(\Delta\chi))^2}{\sin^2(\Delta\chi)\cos^2(el_i)}\left(\sigma_i^{el}\right)^2 \right\} ,
$$

$$
\left(\sigma_i^{rv,oop}\right)^2 = u_\perp^2 \qquad\qquad [\text{A.13}]
$$

$$
\cdot \left\{ \frac{(\sin(el_{i'}) - \sin(el_i)\cos(\Delta\chi))^2}{\sin^2(\Delta\chi)}\,(\sigma_i^{az})^2 + \frac{\cos^2(\gamma_z)}{\cos^2(el_i)}\left(\sigma_i^{el}\right)^2 \right\} ,
$$

$$\text{cov}(ip,oop) = 2u_\perp u_H \sin(\alpha_j \mp \frac{\Delta\chi}{2} - \gamma_{u_H}) \tag{A.14}$$
$$\cdot \frac{\pm\cos(\gamma_z)}{\sin(\Delta\chi)} (\sin(el_{i'}) - \sin(el_i)\cos(\Delta\chi)) \cdot \left\{ (\sigma_i^{az})^2 - \frac{(\sigma_i^{el})^2}{\cos^2(el_i)} \right\} .$$

A.4. Bias Error Propagation

We can assume the radial velocity estimator to work bias-free, therefore we only have to propagate the angular and out-of-plane biases as absolute errors to obtain the bias of u_j:

$$\left| bias_{DD}^{single}(u_j) \right| = \sum_{i=1,2} \left| \frac{\partial u_j}{\partial rv_i} \right| |bias_i^{rv}| , \tag{A.15}$$

with

$$|bias_i^{rv}| = \left| bias_i^{rv,ip} \right| + \left| bias_i^{rv,oop} \right| . \tag{A.16}$$

We find that

$$\left| bias_i^{rv,ip} \right| = u_H \left| \sin\left(\alpha_j \mp \frac{\Delta\chi}{2} - \gamma_{u_H} \right) \right| \tag{A.17}$$
$$\cdot \left\{ |\cos(\gamma_z)| \, |bias_i^{az}| + \frac{|\sin(el_{i'}) - \sin(el_i)\cos(\Delta\chi)|}{|\sin(\Delta\chi)\cos(el_i)|} \left| bias_i^{el} \right| \right\} .$$

Accordingly, the bias or absolute out-of-plane error is given by

$$\left| bias_i^{rv,oop} \right| = |u_\perp| \tag{A.18}$$
$$\cdot \left\{ \frac{|\sin(el_{i'}) - \sin(el_i)\cos(\Delta\chi)|}{|\sin(\Delta\chi)|} |bias_i^{az}| + \frac{|\cos(\gamma_z)|}{|\cos(el_i)|} \left| bias_i^{el} \right| \right\} .$$

A.5. Generalization to Scanning Beam Retrievals

For scanning beams, a retrieval is made using all radial velocity estimates inside one grid cell of the scanning plane, which were recorded
in the time interval T_0. As before, we assume the desired beam angles
to define the lidar plane. Therefore, all single lidar errors appear in the
radial velocity estimates in the vector $\mathbf{b}$ of Eq. 3.24.

Inside on grid cell and during time T_0, we can assume the single lidar
statistical errors and biases on the velocity estimates to be constant for
each lidar. They can be computed using the formulas given above.

Using the notation from Chap. 3.3, we find for the statistical error in the
u_j-component of the wind field in the evaluation plane:

$$[\sigma_{DD}^{\text{single}}(u_j)]^2 = \sum_{n,L1} g_n^2 \left(\hat{\mathbf{e}}_j \cdot \mathbf{M}^{-1} \cdot \hat{\mathbf{r}}_n\right)^2 (\sigma_1^{rv})^2 + \sum_{n,L2} g_n^2 \left(\hat{\mathbf{e}}_j \cdot \mathbf{M}^{-1} \cdot \hat{\mathbf{r}}_n\right)^2 (\sigma_2^{rv})^2 ,$$

$$[\text{A.19}]$$

where the $L1$ and $L2$ sums stand for sums over data taken at lidars one
or two, respectively.

The absolute error is given by

$$\left|bias_{DD}^{\text{single}}(u_j)\right| = \sum_{n,L1} g_n \left|\hat{\mathbf{e}}_j \cdot \mathbf{M}^{-1} \cdot \hat{\mathbf{r}}_n\right| |bias_1^{rv}| + \sum_{n,L2} g_n \left|\hat{\mathbf{e}}_j \cdot \mathbf{M}^{-1} \cdot \hat{\mathbf{r}}_n\right| |bias_2^{rv}| .$$

$$[\text{A.20}]$$

As a rule of thumb, the statistical variance in the scanning beam method
is approximately given by the statistical variance in the intersecting beam
method, divided be half the number of velocity estimates that enter into
the matrix Eq. 3.24. This is due to the fact that, in the scanning beam
method, one grid cell usually contains more than one velocity estimate
per lidar, which leads to reduced statistical uncertainty. For the approximation one should use average angle values inside the grid cell.

$$[\sigma_{DD}^{\text{single}}(u_j, \text{scanning})]^2 \approx \frac{[\sigma_{DD}^{\text{single}}(u_j, \text{intersect.})]^2}{N/2} . \qquad [\text{A.21}]$$

On the other hand, the bias can be approximated by the intersecting beam bias, since it does not scale with the number of data points:

$$\left| bias_{DD}^{single}(u_j, \text{scanning}) \right| \approx \left| bias_{DD}^{single}(u_j, \text{intersect.}) \right| . \qquad [\text{A.22}]$$

B. Lidar Simulation Parameters

The steering parameters used for lidar simulation match those used in the set-up of the LMCT Doppler lidars of the WindTracer type (Chap. 3.1.2). Tab. B.1 summarizes the virtual lidar control parameters. From those parameters, the position and range of each range gate can be computed:

The sampling rate (SR) is the frequency with which the detector records the backscattered signals of the outgoing laser pulses. Each sample has therefore a duration of 1/SR. The number of samples per gate (SpG) then determines the full length of one range gate in time domain:

$$\Delta p_t = \frac{\text{SpG}}{\text{SR}} \, , \tag{B.1}$$

which corresponds to a spatial length of

$$\Delta p = \frac{\text{SpG}}{2\,\text{SR}} \cdot c \, , \tag{B.2}$$

where c is the speed of light.

The range gates are distributed evenly along the lidar beam, starting at the offset range (OR) and ending at the maximum distance (MaxD). Thereby, the distance between range gate centers is given by

$$\text{RGdist} = \frac{\text{MaxD} - \Delta p}{\text{RGnum} - 1} \, , \tag{B.3}$$

where RGnum is the range gate number. This means that the position of the nth range gate centers is located at

$$r_0(n) = \text{OR} + \frac{\Delta p}{2} + (n-1) \cdot \text{RGdist} \quad , \; n = 1, \ldots, \text{RGnum} \, . \tag{B.4}$$

Lidar	Parameter	$u_G = 0$ m/s	$u_G = 5$ m/s	$u_G = 10$ m/s	$u_G = 15$ m/s
Lidar 1	Sampling Rate [Hz]	$2.5 \cdot 10^8$	$2.5 \cdot 10^8$	$2.5 \cdot 10^8$	$2.5 \cdot 10^8$
	Samples per Gate	100	100	110	128
	Offset [m]	350	350	350	350
	Maximum Distance [m]	5520	5520	5520	5520
	Range Gate Number	110	110	110	110
	Pulse Width [s]	$3.0 \cdot 10^{-7}$	$3.0 \cdot 10^{-7}$	$3.0 \cdot 10^{-7}$	$3.0 \cdot 10^{-7}$
	Measurement Frequency [Hz]	10	10	10	10
	Pulse Percentile	20	20	20	20
	Azimuth Range [°]	[315,45]	[315,45]	[315,45]	[315,45]
	Angular Velocity [°/s]	6.2	6.2	6.9	7.9
	Elevation Range [°]	[0,0]	[0,0]	[0,0]	[0,0]
	Angular Velocity EL[°/s]	0	0	0	0
	Position (x,y,z) [m]	[2500,0,10]	[2500,0,10]	[2500,0,10]	[2500,0,10]
Lidar 2	Sampling Rate [Hz]	$2.5 \cdot 10^8$	$2.5 \cdot 10^8$	$2.5 \cdot 10^8$	$2.5 \cdot 10^8$
	Samples per Gate	100	100	110	128
	Offset [m]	350	350	350	350
	Maximum Distance [m]	5520	5520	5520	5520
	Range Gate Number	110	110	110	110
	Pulse Width [s]	$3.7 \cdot 10^{-7}$	$3.7 \cdot 10^{-7}$	$3.7 \cdot 10^{-7}$	$3.7 \cdot 10^{-7}$
	Measurement Frequency [Hz]	10	10	10	10
	Pulse Percentile	20	20	20	20
	Azimuth Range [°]	[225,315]	[225,315]	[225,315]	[225,315]
	Angular Velocity AZ[°/s]	6.2	6.2	6.9	7.9
	Elevation Range [°]	[0,0]	[0,0]	[0,0]	[0,0]
	Angular Velocity EL[°/s]	0	0	0	0
	Position (x,y,z) [m]	[5000,2500,10]	[5000,2500,10]	[5000,2500,10]	[5000,2500,10]

Table B.1.: Control parameters of the virtual lidar measurements used in this study.

The overlap (OL) of adjacent range gates is given by

$$OL = \Delta p - \text{RGdist} .$$

[B.5]

If the overlap is negative, there are gaps between the range gates. The angular velocity ω and the measurement frequency determine the angle $\Delta\beta$ which is scanned by the lidar per mesurement:

$$\Delta\beta = \frac{\omega}{f} .$$

[B.6]

Depending on the distance r along the beam, this angle translates into a circular arc of length

$$\Delta s = \frac{r \cdot \Delta\beta}{180°} .$$

[B.7]

C. Comparative LES and Retrieval Spectra of v

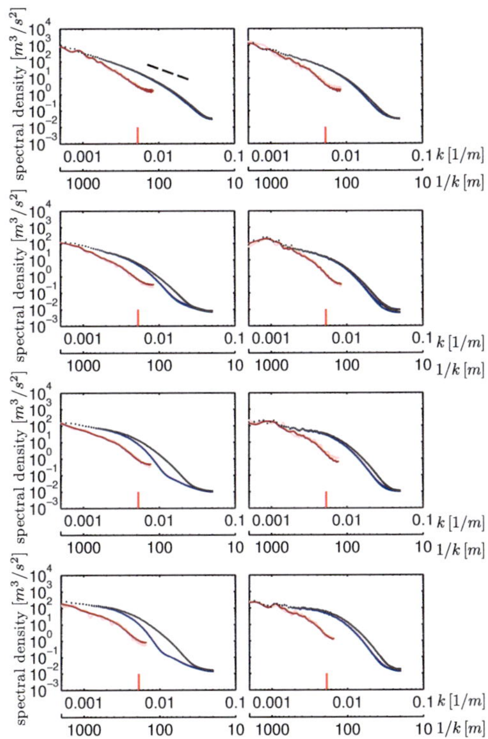

Figure C.1.: Comparative spectral density of the v-component along (left) and across (right) the mean wind direction for the four data sets (u_G increasing from top to bottom). The spectra are shown for the retrieval results (red), the time-averaged LES results (blue) and the LES results (black). The dashed line indicates the slope of $k^{-5/3}$. The pale red lines show a random choice of ten retrieval spectra. The red mark on the k-axis indicated the effective resolution of the simulation and retrieval.

D. Effect of Spatial Smoothing on v-Spectra

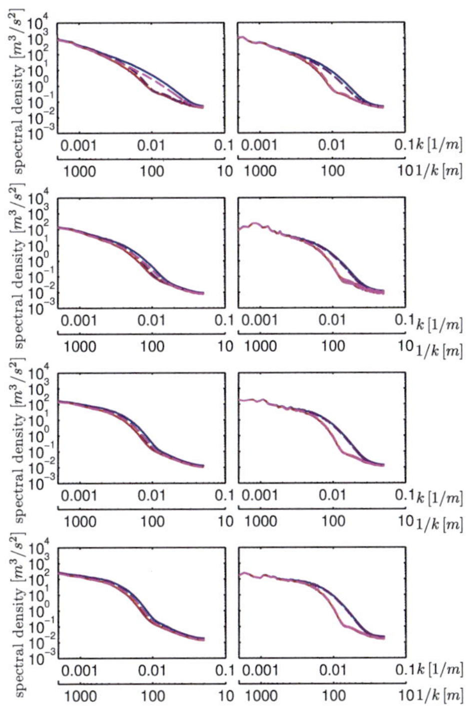

Figure D.1.: Effect of spatial smoothing on the spectral density of the v-component along (left) and across (right) the mean wind direction for the four data sets (u_G increasing from top to bottom). The mean spectra are shown for the time-averaged LES results after applying a moving average filter with the span (Δ_x, Δ_y) in x- and y-direction, respectively: $(\Delta_x, \Delta_y) = (0\text{ m}, 0\text{ m})$ (blue), $(\Delta_x, \Delta_y) = (\Delta, 0\text{ m})$ (dark purple), $(\Delta_x, \Delta_y) = (0\text{ m}, \Delta)$ (light purple), and $(\Delta_x, \Delta_y) = (\Delta, \Delta)$ (red). $\Delta = \{70\text{ m}, 70\text{ m}, 70\text{ m}, 90\text{ m}\}$ for $u_G = \{0\text{ m/s}, 5\text{ m/s}, 10\text{ m/s}, 15\text{ m/s}\}$.

E. Data Sets for Spatial Scale Analysis

Data Set	Field	Direction	Data Set Size at Wind Speed				Data Loss Count at Wind Speed			
			0 m/s	5 m/s	10 m/s	15 m/s	0 m/s	5 m/s	10 m/s	15 m/s
LES	u	x			1800		0	0	4	25
LES	u	y			1800		0	0	0	0
LES	u	t			2704		533	9	2	1
LES	v	x			1800		40	0	10	0
LES	v	y			1800		0	0	0	0
LES	v	t			2704		542	23	2	0
LESAVG	u	x	124	124	138	158	0	0	1	2
LESAVG	u	y	124	124	138	158	0	0	0	0
LESAVG	v	x	124	124	138	158	2	0	1	0
LESAVG	v	y	124	124	138	158	0	0	0	0
RET	u	x	124	124	138	158	0	1	1	0
RET	u	y	124	124	138	158	0	0	0	0
RET	v	x	124	124	138	158	0	0	1	6
RET	v	y	124	124	138	158	0	0	0	0

Table E.1.: Statistics of the data set used for correlation length computation. In x- and y-direction, the correlation lengths are computed for each time step. Virtual towers on a 100 m grid were used to compute the correlation length from time series. Data loss occurred when the autocorrelation function did not exhibit a zero-crossing in the desired direction.

Data Set	Field	Direction	Data Set Size at Wind Speed				Data Loss Count at Wind Speed			
			0 m/s	5 m/s	10 m/s	15 m/s	0 m/s	5 m/s	10 m/s	15 m/s
LES	u	x			1800		0	10	85	134
LES	u	y			1800		272	0	0	0
LES	u	t			441		127	30	27	19
LES	v	x			1800		164	74	116	42
LES	v	y			1800		0	0	0	0
LES	v	t			441		110	83	30	19
LESAVG	u	x	620	620	690	790	0	24	15	15
LESAVG	u	y	620	620	690	790	106	0	0	0
LESAVG	v	x	620	620	690	790	74	89	24	5
LESAVG	v	y	620	620	690	790	17	0	0	0
RET	u	x	620	620	690	790	82	94	131	142
RET	u	y	620	620	690	790	154	18	4	11
RET	v	x	620	620	690	790	206	159	139	101
RET	v	y	620	620	690	790	50	25	22	22

Table E.2.: Statistics of the data set used for wavelet analysis. For each LES time step, one series in x-direction at random y-position was analyzed, and vice versa, for each field. For LESAVG and RET data, five random series were used per time step. Virtual towers positioned on a 250 m grid were used for the time series analysis. Data loss occurred when a series did not exhibit a $\tilde{E}_1$ maximum in the scale range (cf. Chap. 6.2.4).

Data Set	Field	Dir	Total Structure Count at Wind Speed and Levels									
			0 m/s					5 m/s				
			0	-0.2	-0.4	-0.6	-0.8	0	-0.2	-0.4	-0.6	-0.8
LES	u	x	9933	8765	6994	4984	3111	12190	9995	7153	4591	2726
LES	u	y	6281	5380	4079	2889	1944	49097	38480	23511	11537	4748
LES	u	t	1580	1345	988	667	438	3423	2824	1977	1218	698
LES	v	x	7042	5967	4457	3055	2030	10266	8267	5836	3769	2394
LES	v	y	10490	9090	7013	4862	2954	13047	11659	9269	6049	3313
LES	v	t	1759	1514	1169	760	466	2504	2046	1410	904	550
LESAVG	u	x	3220	2875	2297	1674	1045	3788	3148	2280	1514	928
LESAVG	u	y	2126	1826	1394	932	633	16586	13082	8079	3902	1650
LESAVG	v	x	2383	2022	1499	1037	687	3138	2578	1839	1140	748
LESAVG	v	y	3408	2974	2325	1601	970	4218	3776	3058	1997	1122
RET	u	x	1290	1164	916	700	507	1596	1381	1072	780	574
RET	u	y	1105	982	775	572	430	4266	2828	1862	1208	751
RET	v	x	1009	885	712	551	421	1236	1057	844	654	491
RET	v	y	1578	1454	1251	997	735	2250	2134	1803	1299	802
			10 m/s					15 m/s				
			0	-0.2	-0.4	-0.6	-0.8	0	-0.2	-0.4	-0.6	-0.8
LES	u	x	10182	8598	6377	4210	2575	9960	8425	6236	4158	2545
LES	u	y	50446	42092	28090	14135	5412	49629	41766	28211	14534	5749
LES	u	t	4822	4010	2776	1645	857	5935	4903	3349	1900	929
LES	v	x	16097	12783	8377	4993	2768	18998	15375	10185	5810	3112
LES	v	y	27949	23578	16659	9519	4454	38090	32521	22425	12093	5149
LES	v	t	7046	5376	3152	1644	809	12717	9728	5631	2641	1109
LESAVG	u	x	3638	3081	2281	1553	961	4012	3482	2652	1795	1115
LESAVG	u	y	18818	15749	10861	5548	2154	21121	17928	12258	6418	2529
LESAVG	v	x	4964	4111	2828	1791	1064	6165	5123	3571	2208	1278
LESAVG	v	y	8717	7328	5269	3143	1578	14482	12401	8751	4812	2116
RET	u	x	1498	1277	975	762	593	1584	1328	1042	784	603
RET	u	y	5969	3636	2150	1329	768	6378	3622	1929	1138	677
RET	v	x	1802	1540	1187	857	613	2150	1884	1470	1054	729
RET	v	y	2570	2276	1770	1306	921	3266	2839	2202	1552	1034

Table E.3.: Total number of detected wavelet ramps for the data sets.

Data Set	Field	Direction	Data Set Size at Wind Speed			
			0 m/s	5 m/s	10 m/s	15 m/s
LES	u	x			100	
LES	u	y			100	
LES	u	t			500	
LES	v	x			100	
LES	v	y			100	
LES	v	t			500	
LESAVG	u	x	124	124	138	158
LESAVG	u	y	124	124	138	158
LESAVG	v	x	124	124	138	158
LESAVG	v	y	124	124	138	158
RET	u	x	124	124	138	158
RET	u	y	124	124	138	158
RET	v	x	124	124	138	158
RET	v	y	124	124	138	158

Table E.4.: Statistics of the data set used for cluster analysis. For 100 random LES time steps, and all RET and LESAVG time steps, clusters were computed for all fields in 2D. Time series at 500 virtual towers, evenly distributed across the area, were used to compute 1D clusters in time direction. No data loss occurred.

Data Set	Field	Dir	Total Structure Count at Wind Speed and Levels											
			0 m/s						5 m/s					
			-3	-2.5	-2	-1.5	-1	-0.5	-3	-2.5	-2	-1.5	-1	-0.5
LES	u	x/y	521	3027	9807	17969	23268	23446	1381	8531	25043	37333	33176	28917
LES	u	t	31	133	558	1470	2438	3145	42	203	942	2430	4175	5845
LES	v	x/y	396	3064	11002	20323	23597	20007	1108	4676	12891	21452	22312	16631
LES	v	t	33	156	539	1443	2461	3273	58	284	960	2272	3873	5126
LESAVG	u	x/y	474	2986	10246	18788	23816	23912	1727	9747	27318	40522	36455	31231
LESAVG	v	x/y	366	2983	11639	21641	24389	22563	1013	4583	12312	20146	20481	13118
RET	u	x/y	16	175	895	1965	2648	2467	42	258	1181	3073	4549	3859
RET	v	x/y	1	96	757	2405	3230	2462	11	273	993	2393	3268	2685
			10 m/s						15 m/s					
			-3	-2.5	-2	-1.5	-1	-0.5	-3	-2.5	-2	-1.5	-1	-0.5
LES	u	x/y	198	5255	21140	32984	32322	29504	96	3117	17727	33955	34404	31018
LES	u	t	10	213	1239	3582	6267	8680	3	113	1300	4446	8613	11995
LES	v	x/y	2217	7059	16911	29894	34664	29591	2615	8001	19702	35926	45622	39877
LES	v	t	178	627	1808	4376	7831	11453	255	908	2728	6426	11854	17399
LESAVG	u	x/y	382	6942	24241	36908	34849	32165	143	4520	22841	41306	39835	34037
LESAVG	v	x/y	2120	6655	15696	26313	28396	20465	2692	8179	19900	33691	39369	30846
RET	u	x/y	1	122	898	2645	4205	3799	4	83	837	2446	3606	3676
RET	v	x/y	65	363	1379	2796	3499	3542	77	330	1229	2667	3879	4147

Table E.5.: Total number of detected clusters for the data sets.

F. Spatial Autocorrelation Results

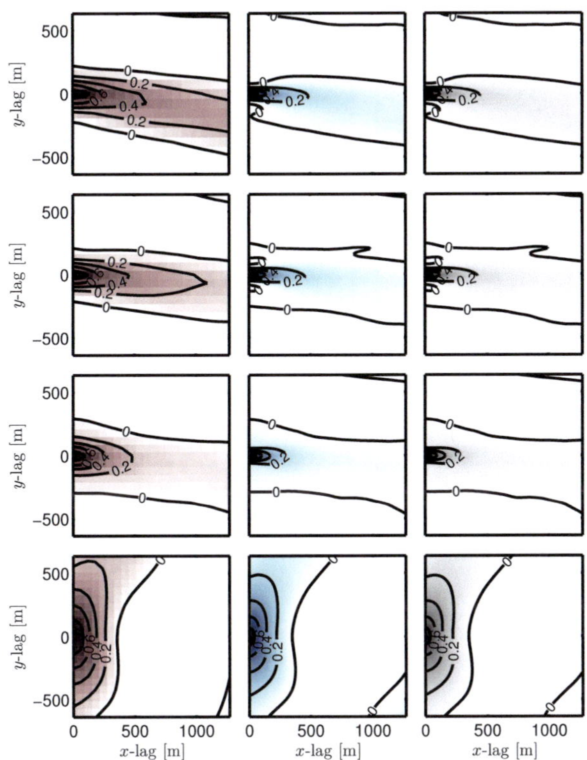

Figure F.1.: Zoom into the time-averaged spatial autocorrelations of the u-components of the wind fields for $u_G = \{0\ \text{m/s},\ 5\ \text{m/s},\ 10\ \text{m/s},\ 15\ \text{m/s}\}$ (bottom to top) for the three data sets retrieval results (red), time-averaged LES results (blue), and full-resolution LES results (black).

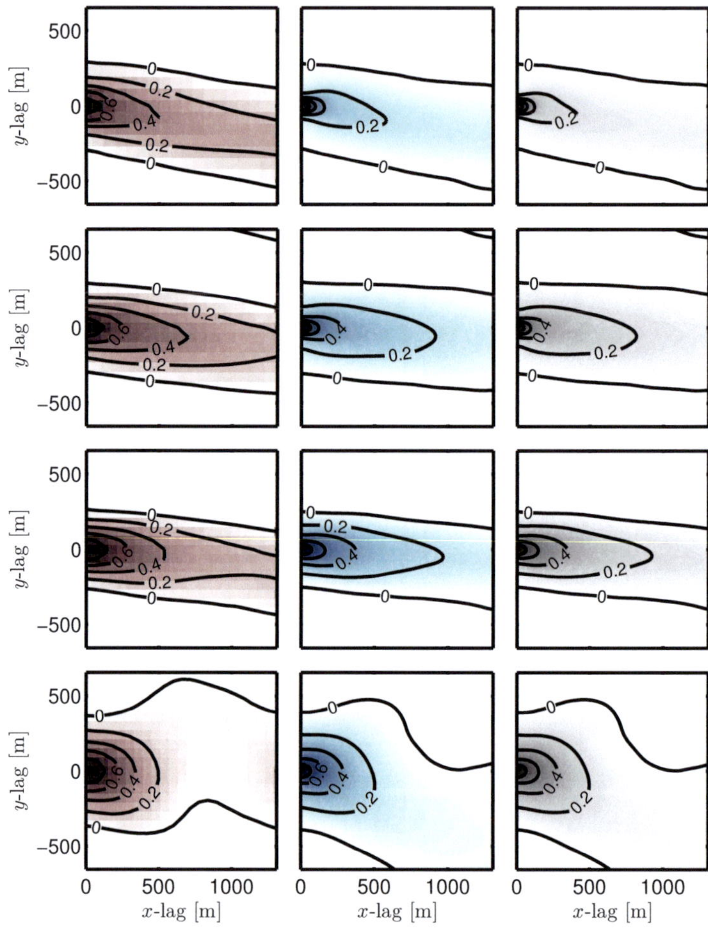

Figure F.2.: Zoom into time-averaged spatial autocorrelations of the v-components of the wind fields for $u_G = \{0\ \text{m/s}, 5\ \text{m/s}, 10\ \text{m/s}, 15\ \text{m/s}\}$ (bottom to top) for the three data sets retrieval results (red), time-averaged LES results (blue), and full-resolution LES results (black).

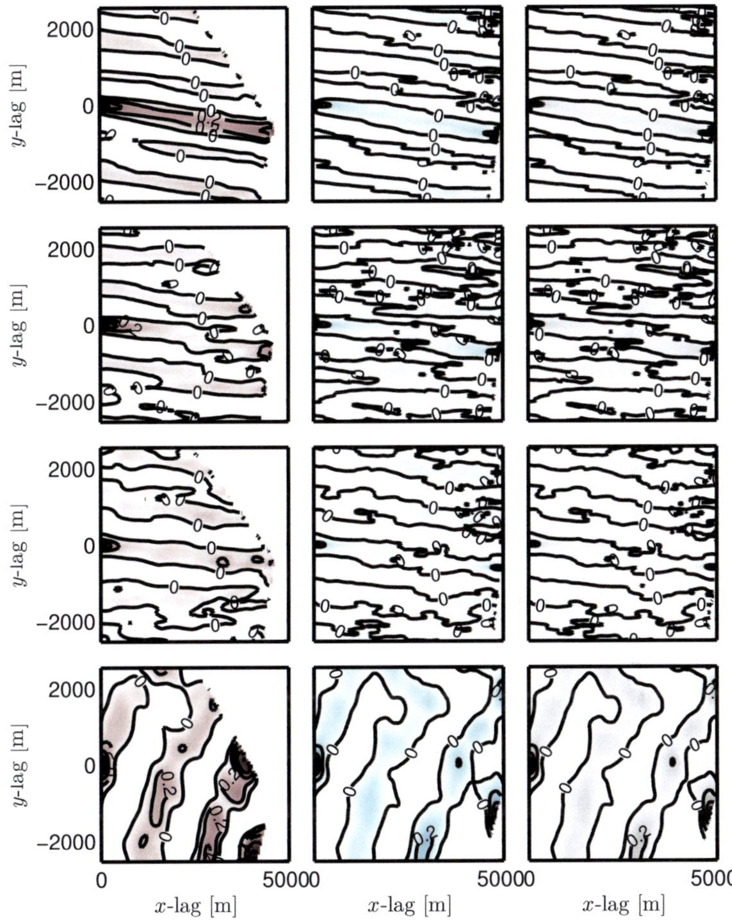

Figure F.3.: Full-range time-averaged spatial autocorrelations of the u-components of the wind fields for $u_G = \{0\text{ m/s},\ 5\text{ m/s},\ 10\text{ m/s},\ 15\text{ m/s}\}$ (bottom to top) for the three data sets retrieval results (red), time-averaged LES results (blue), and full-resolution LES results (black).

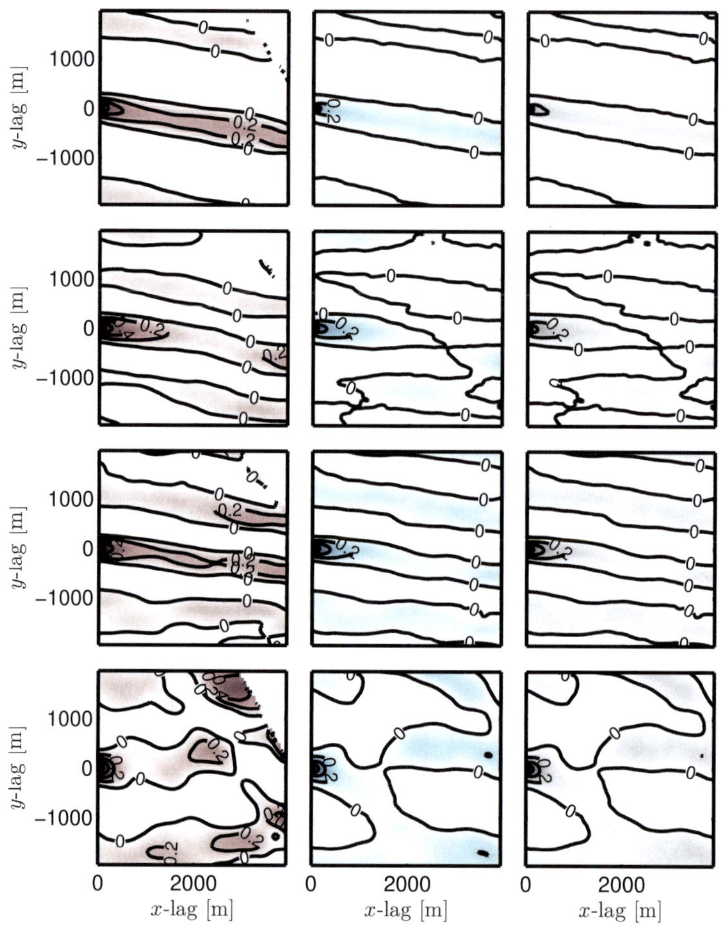

Figure F.4.: Full-range time-averaged spatial autocorrelations of the v-components of the wind fields for $u_G = \{0\text{ m/s}, 5\text{ m/s}, 10\text{ m/s}, 15\text{ m/s}\}$ (bottom to top) for the three data sets retrieval results (red), time-averaged LES results (blue), and full-resolution LES results (black).

G. Wavelet Length Scales for Varying Cutoff Values

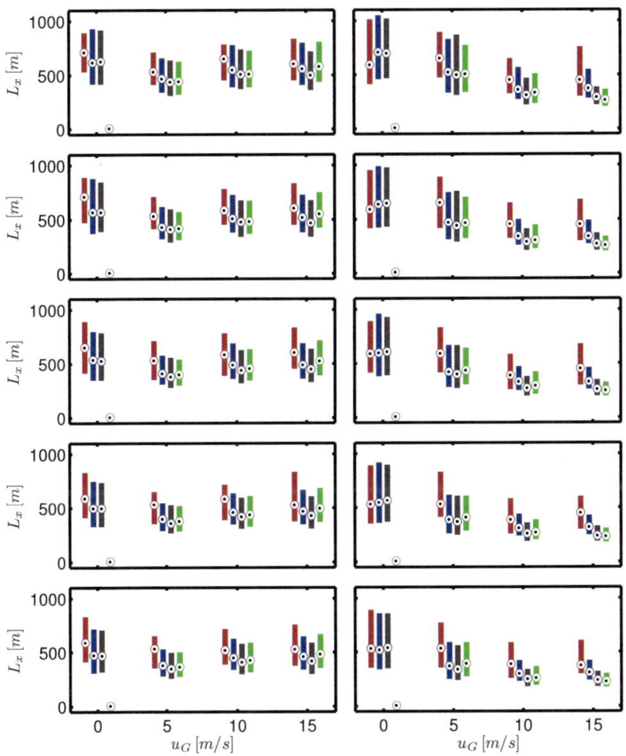

Figure G.1.: Wavelet ramp lengths in x direction for varying cutoff values $K = \{0, 0.2, 0.4, 0.6, 0.8\}$ (bottom to top) for the wind fields components u (left) and v (right). Colors and range as in Fig 6.17.

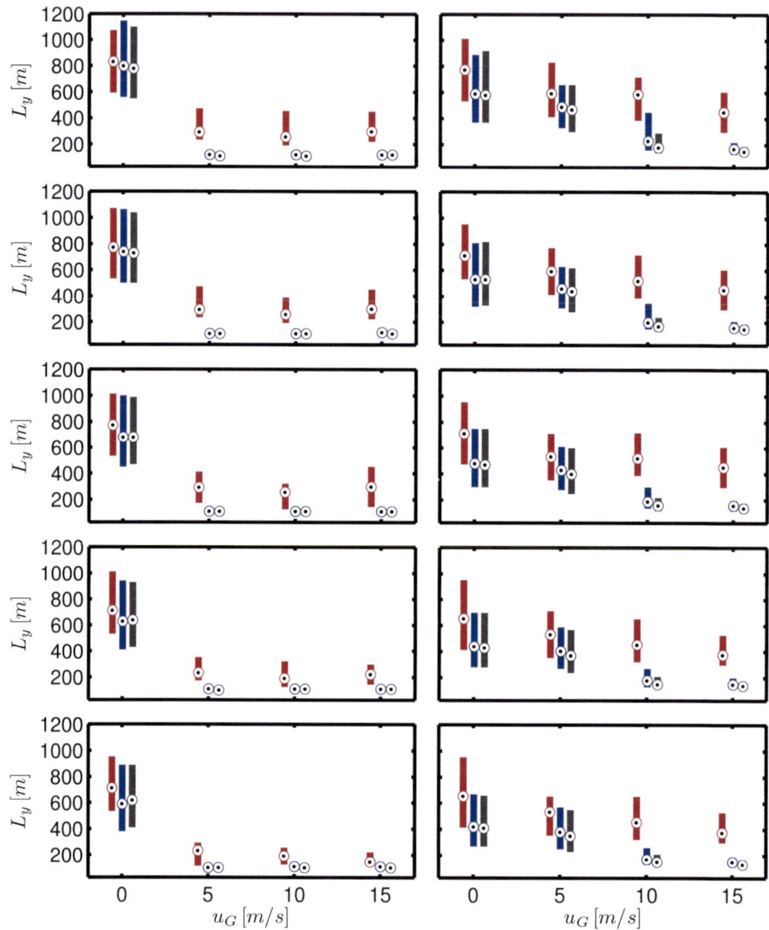

Figure G.2.: Wavelet ramp lengths in y direction for varying cutoff values $K = \{0, 0.2, 0.4, 0.6, 0.8\}$ (bottom to top) for the wind fields components u (left) and v (right). Colors and range as in Fig 6.17.

H. Clustering Length Scales for Varying Cutoff Values

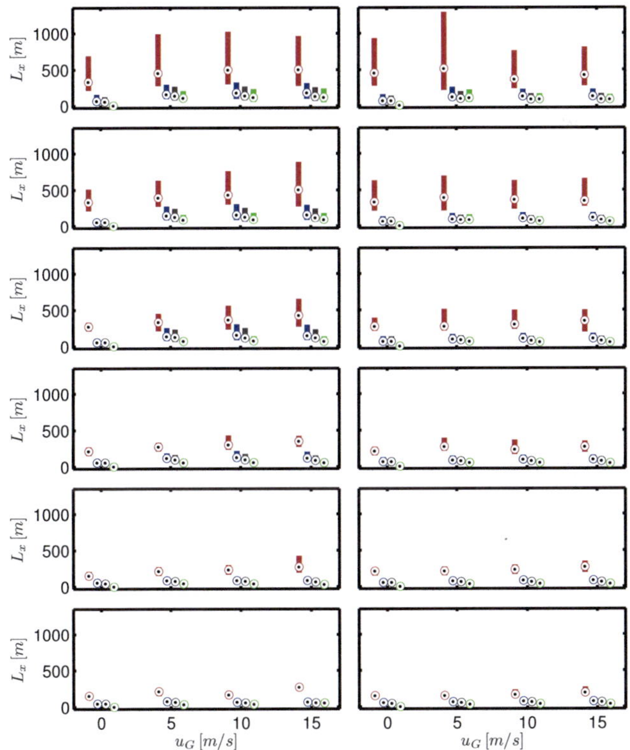

Figure H.1.: Cluster lengths in x direction for varying cutoff values $\sigma \cdot \{-3, -2.5, -2, -1.5, -1, -0.5\}$ (bottom to top) for the wind fields components u (left) and v (right). Colors and range as in Fig 6.17.

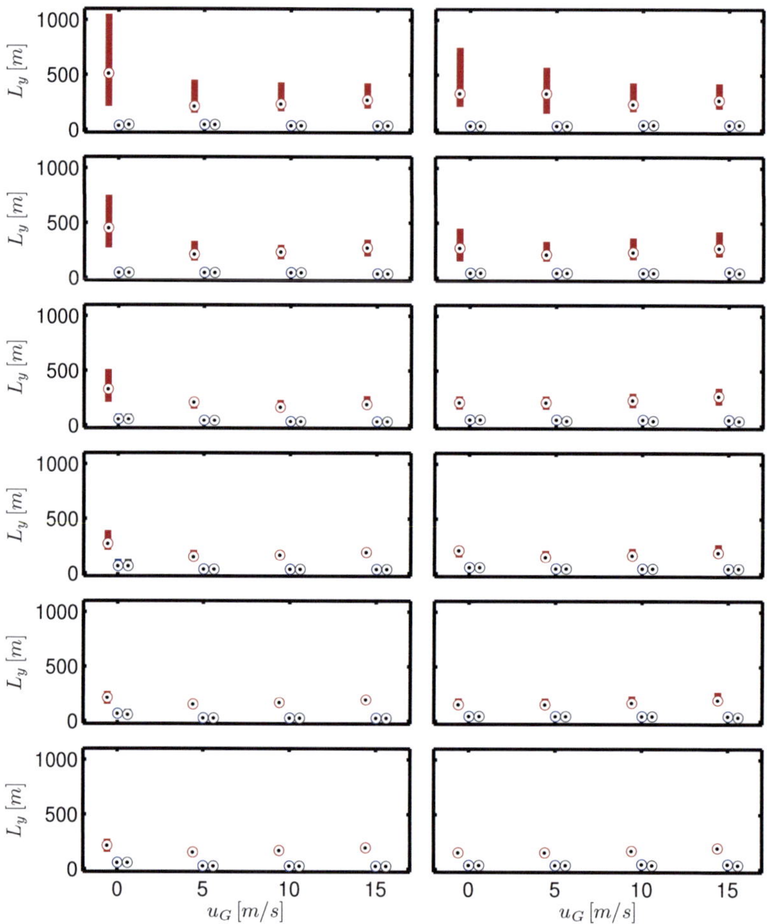

Figure H.2.: Cluster lengths in y direction for varying cutoff values $\sigma \cdot \{-3, -2.5, -2, -1.5, -1, -0.5\}$ (bottom to top) for the wind fields components u (left) and v (right). Colors and range as in Fig 6.17.

I. Data Sets for Vertical Wind Field Analysis

fit y to w		0 m/s					5 m/s				
$y = m \cdot \sigma + b$	steps $[T_0]$	m [m]			b [m/s]		m [m]			b [m/s]	
$w_{LESAVG,intp}$, σ_{RET}	1	8.77	±	7e-02	-0.08	± 1e-03	7.29	±	6e-02	-0.04	± 4e-04
	2	11.89	±	1e-01	-0.07	± 1e-03	10.12	±	8e-02	-0.04	± 4e-04
	3	13.46	±	1e-01	-0.06	± 1e-03	11.69	±	9e-02	-0.04	± 4e-04
	5	14.91	±	1e-01	-0.04	± 1e-03	13.44	±	1e-01	-0.03	± 4e-04
	8	15.62	±	1e-01	-0.02	± 1e-03	14.86	±	1e-01	-0.03	± 4e-04
	13	15.87	±	2e-01	0.01	± 1e-03	16.19	±	2e-01	-0.03	± 4e-04
$w_{LESAVG,intp,sm}$, σ_{RET}	1	8.02	±	4e-02	-0.07	± 6e-04	6.65	±	3e-02	-0.04	± 3e-04
	2	10.89	±	5e-02	-0.06	± 6e-04	9.23	±	5e-02	-0.04	± 3e-04
	3	12.33	±	6e-02	-0.05	± 6e-04	10.68	±	5e-02	-0.03	± 3e-04
	5	13.68	±	7e-02	-0.04	± 5e-04	12.31	±	6e-02	-0.03	± 3e-04
	8	14.35	±	7e-02	-0.02	± 5e-04	13.66	±	8e-02	-0.03	± 3e-04
	13	14.60	±	8e-02	0.01	± 5e-04	14.97	±	9e-02	-0.02	± 2e-04
w_{LESAVG}, σ_{LESAVG}	1	8.86	±	3e-03	-0.15	± 1e-04	5.02	±	2e-03	-0.09	± 5e-05
	2	12.10	±	4e-03	-0.09	± 1e-04	7.07	±	3e-03	-0.08	± 5e-05
	3	13.80	±	5e-03	-0.04	± 1e-04	8.26	±	4e-03	-0.08	± 6e-05
	5	15.53	±	7e-03	0.03	± 1e-04	9.65	±	5e-03	-0.07	± 6e-05
	8	16.73	±	1e-02	0.09	± 2e-04	10.83	±	7e-03	-0.05	± 6e-05
	13	17.79	±	1e-02	0.15	± 2e-04	12.09	±	1e-02	-0.04	± 6e-05
w_{LES}, σ_{LES}	1	14.71	±	4e-03	-0.17	± 9e-05	104.74	±	4e-02	-0.09	± 4e-05
	2	16.32	±	4e-03	-0.12	± 9e-05	58.49	±	2e-02	-0.08	± 4e-05
	3	17.43	±	5e-03	-0.07	± 1e-04	40.35	±	2e-02	-0.07	± 5e-05
	5	18.86	±	7e-03	0.00	± 1e-04	25.63	±	1e-02	-0.06	± 5e-05
	8	20.16	±	9e-03	0.08	± 1e-04	17.01	±	1e-02	-0.05	± 5e-05
	13	21.51	±	1e-02	0.15	± 1e-04	11.27	±	9e-03	-0.03	± 5e-05

Table I.1.: Fit parameters for the linear fit of the FTLE coefficients to the vertical wind field, Part I: $u_G = 0$ m/s and $u_G = 5$ m/s

fit y to w		10 m/s						15 m/s						
$y = m \cdot \sigma + b$	steps $[T_0]$	m [m]			b [m/s]			m [m]			b [m/s]			
$w_{\text{LESAVG,intp}}$, σ_{RET}	1	5.86	$\pm$	7e-02	-0.04	$\pm$	6e-04	5.43	$\pm$	9e-02	-0.04	$\pm$	9e-04	
	2	8.25	$\pm$	9e-02	-0.04	$\pm$	6e-04	7.78	$\pm$	1e-01	-0.04	$\pm$	9e-04	
	3	9.56	$\pm$	1e-01	-0.04	$\pm$	6e-04	8.98	$\pm$	1e-01	-0.04	$\pm$	9e-04	
	5	10.93	$\pm$	1e-01	-0.04	$\pm$	6e-04	10.04	$\pm$	2e-01	-0.04	$\pm$	9e-04	
	8	11.89	$\pm$	2e-01	-0.03	$\pm$	6e-04	10.58	$\pm$	2e-01	-0.03	$\pm$	9e-04	
	13	12.77	$\pm$	2e-01	-0.03	$\pm$	6e-04	11.14	$\pm$	3e-01	-0.03	$\pm$	9e-04	
$w_{\text{LESAVG,intp,sm}}$, σ_{RET}	1	5.50	$\pm$	4e-02	-0.04	$\pm$	4e-04	4.87	$\pm$	4e-02	-0.04	$\pm$	5e-04	
	2	7.75	$\pm$	6e-02	-0.04	$\pm$	4e-04	6.95	$\pm$	6e-02	-0.04	$\pm$	5e-04	
	3	8.99	$\pm$	7e-02	-0.04	$\pm$	4e-04	8.05	$\pm$	7e-02	-0.04	$\pm$	5e-04	
	5	10.32	$\pm$	8e-02	-0.04	$\pm$	4e-04	9.07	$\pm$	9e-02	-0.04	$\pm$	5e-04	
	8	11.30	$\pm$	1e-01	-0.03	$\pm$	4e-04	9.61	$\pm$	1e-01	-0.03	$\pm$	5e-04	
	13	12.22	$\pm$	1e-01	-0.03	$\pm$	4e-04	10.18	$\pm$	1e-01	-0.03	$\pm$	4e-04	
w_{LESAVG} , σ_{LESAVG}	1	4.29	$\pm$	2e-03	-0.12	$\pm$	8e-05	4.27	$\pm$	2e-03	-0.16	$\pm$	1e-04	
	2	6.06	$\pm$	3e-03	-0.12	$\pm$	8e-05	6.04	$\pm$	3e-03	-0.15	$\pm$	1e-04	
	3	7.06	$\pm$	4e-03	-0.11	$\pm$	8e-05	7.01	$\pm$	4e-03	-0.14	$\pm$	1e-04	
	5	8.16	$\pm$	6e-03	-0.09	$\pm$	9e-05	8.06	$\pm$	6e-03	-0.12	$\pm$	1e-04	
	8	9.05	$\pm$	8e-03	-0.07	$\pm$	9e-05	8.88	$\pm$	8e-03	-0.09	$\pm$	1e-04	
	13	10.01	$\pm$	1e-02	-0.06	$\pm$	9e-05	9.77	$\pm$	1e-02	-0.07	$\pm$	1e-04	
w_{LES} , σ_{LES}	1	92.27	$\pm$	4e-02	-0.12	$\pm$	7e-05	92.32	$\pm$	4e-02	-0.17	$\pm$	9e-05	
	2	48.93	$\pm$	2e-02	-0.11	$\pm$	7e-05	46.39	$\pm$	2e-02	-0.14	$\pm$	1e-04	
	3	33.17	$\pm$	2e-02	-0.10	$\pm$	8e-05	31.47	$\pm$	2e-02	-0.13	$\pm$	1e-04	
	5	20.10	$\pm$	1e-02	-0.08	$\pm$	8e-05	18.65	$\pm$	1e-02	-0.10	$\pm$	1e-04	
	8	12.82	$\pm$	1e-02	-0.06	$\pm$	8e-05	11.78	$\pm$	1e-02	-0.07	$\pm$	1e-04	
	13	8.31	$\pm$	1e-02	-0.04	$\pm$	8e-05	7.63	$\pm$	1e-02	-0.05	$\pm$	1e-04	

Table I.2.: Fit parameters for the linear fit of the FTLE coefficients to the vertical wind field, Part II: $u_G = 10$ m/s and $u_G = 15$ m/s

Contingency Table: Agreement [%]		steps $[T_0]$	0 m/s w		5 m/s w		10 m/s w		15 m/s w	
			≥ 0	< 0	≥ 0	< 0	≥ 0	< 0	≥ 0	< 0
$\sigma_{RET},\ w_{LESAVG,intp}$	≥ 0	1	41	30	38	45	41	47	44	47
	< 0		7	22	3	14	3	9	3	7
	≥ 0	2	40	29	38	44	41	47	43	46
	< 0		8	23	3	15	3	9	3	7
	≥ 0	3	39	28	38	44	41	46	43	46
	< 0		9	24	3	15	3	9	3	8
	≥ 0	5	37	25	37	43	41	46	43	46
	< 0		11	27	4	16	3	10	4	8
	≥ 0	8	34	23	37	42	40	45	43	45
	< 0		14	29	4	17	4	10	4	8
	≥ 0	13	30	19	36	40	40	45	42	45
	< 0		18	33	5	19	4	11	4	9
$\sigma_{RET},\ w_{LESAVG,intp,sm}$	≥ 0	1	46	25	41	41	44	45	46	44
	< 0		5	24	2	15	2	9	2	8
	≥ 0	2	45	24	41	41	44	44	46	43
	< 0		6	25	2	16	2	10	2	8
	≥ 0	3	44	22	41	40	44	44	46	43
	< 0		7	26	2	16	2	10	2	8
	≥ 0	5	42	20	41	39	43	44	46	43
	< 0		9	29	3	17	3	10	3	9
	≥ 0	8	39	18	40	38	43	43	46	42
	< 0		12	31	3	18	3	11	3	9
	≥ 0	13	35	15	39	37	42	42	45	42
	< 0		16	34	4	20	4	12	4	10
$\sigma_{LESAVG},\ w_{LESAVG}$	≥ 0	1	46	18	41	45	44	48	46	47
	< 0		2	35	0	14	0	7	0	7
	≥ 0	2	43	14	40	44	44	48	46	46
	< 0		4	39	0	15	0	8	0	7
	≥ 0	3	40	11	40	43	44	47	46	46
	< 0		8	42	1	16	0	9	1	8
	≥ 0	5	34	7	39	42	43	46	45	44
	< 0		13	45	1	18	1	10	1	9
	≥ 0	8	27	5	38	39	42	44	44	43
	< 0		20	48	2	20	2	11	2	11
	≥ 0	13	21	3	37	37	41	43	43	42
	< 0		27	50	4	23	3	13	3	11
$\sigma_{LES},\ w_{LES}$	≥ 0	1	46	18	40	45	43	47	45	46
	< 0		2	35	0	15	0	10	0	9
	≥ 0	2	43	14	39	44	43	47	44	45
	< 0		5	39	0	16	1	10	1	10
	≥ 0	3	40	11	39	43	42	46	44	45
	< 0		8	41	1	17	1	11	2	10
	≥ 0	5	34	8	38	41	41	45	43	43
	< 0		14	45	2	19	2	12	3	11
	≥ 0	8	27	5	37	39	40	43	41	42
	< 0		20	48	3	21	3	13	4	13
	≥ 0	13	21	3	35	37	39	42	40	41
	< 0		26	50	5	24	5	15	5	14

Table I.3.: Contingency table: Agreement and disagreement of signs of the vertical wind w and the FTLE, using the appropriate vertical wind fields (cf. Chap. 7.3).

Contigency Table:			0 m/s		5 m/s		10 m/s		15 m/s	
Agreement [%]			w		w		w		w	
		steps $[T_0]$	≥ 0	< 0	≥ 0	< 0	≥ 0	< 0	≥ 0	< 0
DIV_{RET}, $w_{LESAVG,intp}$	≥ 0	1	27	23	22	24	23	25	24	25
	< 0		21	28	18	36	21	31	22	28
DIV_{RET}, $w_{LESAVG,intp,sm}$	≥ 0	1	31	20	25	21	25	23	26	24
	< 0		21	29	19	35	21	31	23	28
DIV_{LESAVG}, w_{LESAVG}	≥ 0	1	42	4	37	6	39	7	41	7
	< 0		5	48	4	53	5	49	5	47
DIV_{LES}, w_{LES}	≥ 0	1	41	5	35	8	37	10	39	10
	< 0		7	48	5	52	6	47	6	45

Table I.4.: Contingency table: Agreement and disagreement of signs of the vertical wind w and w_{DIV}, using the appropriate vertical wind fields (cf. Chap. 7.3).

Danksagung

Die vorliegende Doktorarbeit wurde von 2011 bis 2014 am Institut für Meteorologie und Klimaforschung des Karlsruher Instituts für Technologie (KIT) durchgeführt. Sie erfolgte im Rahmen der KIT Young Investigator Group "Exploring Coherent Structures using Dual-Doppler Lidar Systems" unter der Leitung von Dr. Katja Träumner.

Mein besonderer Dank richtet sich an Prof. Kottmeier für die Betreuung der Arbeit und an Prof. Raasch für die Übernahme des Korreferats. Prof. Kottmeier hat mir jederzeit unterstützend zur Seite gestanden. Durch sein Vertrauen konnte ich meine Arbeit selbständig gestalten und Ergebnisse auf Konferenzen präsentieren. Dank der Zusammenarbeit mit Prof. Raasch und seiner Bereitstellung der LES-Daten ist diese Arbeit erst möglich geworden.

Einen herzlichen Dank möchte ich Dr. Katja Träumner aussprechen. Ich hatte das Glück in ihrer Arbeitsgruppe von ihrer fachlichen Kompetenz, ihrem großen Einsatz und ihrer freundlichen Anleitung profitieren zu können.
Hier möchte ich auch meiner Arbeitsgruppe danken, insbesondere meinem Bürokollegen Thomas Damian: dank euch komme ich jeden Tag gern zur Arbeit.

Darüber hinaus geht mein Dank die Wissenschaftler, Sektretärinnen, Werkstattmitarbeiter und System-Administratoren des IMK-TRO, auf deren Hilfsbereitsschaft und Freundlichkeit ich mich stets verlassen konnte. Besonders möchte ich Dr. Andreas Wieser für seine Hilfe beim Lidar-Betrieb danken. Danke auch an alle, die diese Arbeit gelesen und korrigiert haben.

Vielen Dank an die KIT Graduiertenschule GRACE für die Finanzierung meines Auslandsaufenthaltes, und an Dr. Ron Calhoun und seine Gruppe an der Arizona State University.

Für ihre vielseitige Unterstützung, Zeit und allgemeine Großartigkeit danke ich meinen Freunden, insbesondere Susanne Butz und Jochen Zimmer. Ein besonderer Dank geht an meine Familie für ihren Glauben an mich und ihre Ermutigung.
Und an Philipp Jung, für alles.

Karlsruhe, im Dezember 2013   *Christina Stawiarski*

Wissenschaftliche Berichte des Instituts für Meteorologie und Klimaforschung des Karlsruher Instituts für Technologie (0179-5619)

Bisher erschienen:

Nr. 1: *Fiedler, F. / Prenosil, T.*
Das MESOKLIP-Experiment. (Mesoskaliges Klimaprogramm im Oberrheintal).
August 1980

Nr. 2: *Tangermann-Dlugi, G.*
Numerische Simulationen atmosphärischer Grenzschichtströmungen über langgestreckten mesoskaligen Hügelketten bei neutraler thermischer Schichtung.
August 1982

Nr. 3: *Witte, N.*
Ein numerisches Modell des Wärmehaushalts fließender Gewässer unter Berücksichtigung thermischer Eingriffe.
Dezember 1982

Nr. 4: *Fiedler, F. / Höschele, K. (Hrsg.)*
Prof. Dr. Max Diem zum 70. Geburtstag.
Februar 1983 (vergriffen)

Nr. 5: *Adrian, G.*
Ein Initialisierungsverfahren für numerische mesoskalige Strömungsmodelle.
Juli 1985

Nr. 6: *Dorwarth, G.*
Numerische Berechnung des Druckwiderstandes typischer Geländeformen.
Januar 1986

Nr. 7: *Vogel, B.; Adrian, G. / Fiedler, F.*
MESOKLIP-Analysen der meteorologischen Beobachtungen von mesoskaligen Phänomenen im Oberrheingraben.
November 1987

Nr. 8: *Hugelmann, C.-P.*
Differenzenverfahren zur Behandlung der Advektion.
Februar 1988

Nr. 9:   *Hafner, T.*
Experimentelle Untersuchung zum Druckwiderstand der Alpen.
April 1988

Nr. 10:   *Corsmeier, U.*
Analyse turbulenter Bewegungsvorgänge in der maritimen
atmosphärischen Grenzschicht.
Mai 1988

Nr. 11:   *Walk, O. / Wieringa, J.(eds)*
Tsumeb Studies of the Tropical Boundary-Layer Climate.
Juli 1988

Nr. 12:   *Degrazia, G. A.*
Anwendung von Ähnlichkeitsverfahren auf die turbulente Diffusion
in der konvektiven und stabilen Grenzschicht.
Januar 1989

Nr. 13:   *Schädler, G.*
Numerische Simulationen zur Wechselwirkung zwischen Landober-
flächen und atmophärischer Grenzschicht.
November 1990

Nr. 14:   *Heldt, K.*
Untersuchungen zur Überströmung eines mikroskaligen Hindernisses
in der Atmosphäre.
Juli 1991

Nr. 15:   *Vogel, H.*
Verteilungen reaktiver Luftbeimengungen im Lee einer Stadt –
Numerische Untersuchungen der relevanten Prozesse.
Juli 1991

Nr. 16:   *Höschele, K.(ed.)*
Planning Applications of Urban and Building Climatology – Proceedings
of the IFHP / CIB-Symposium Berlin, October 14-15, 1991.
März 1992

Nr. 17:   *Frank, H. P.*
Grenzschichtstruktur in Fronten.
März 1992

Nr. 18:   *Müller, A.*
Parallelisierung numerischer Verfahren zur Beschreibung von
Ausbreitungs- und chemischen Umwandlungsprozessen in der
atmosphärischen Grenzschicht.
Februar 1996

Nr. 19: *Lenz, C.-J.*
Energieumsetzungen an der Erdoberfläche in gegliedertem Gelände.
Juni 1996

Nr. 20: *Schwartz, A.*
Numerische Simulationen zur Massenbilanz chemisch reaktiver
Substanzen im mesoskaligen Bereich.
November 1996

Nr. 21: *Beheng, K. D.*
Professor Dr. Franz Fiedler zum 60. Geburtstag.
Januar 1998

Nr. 22: *Niemann, V.*
Numerische Simulation turbulenter Scherströmungen mit einem
Kaskadenmodell.
April 1998

Nr. 23: *Koßmann, M.*
Einfluß orographisch induzierter Transportprozesse auf die Struktur
der atmosphärischen Grenzschicht und die Verteilung von
Spurengasen.
April 1998

Nr. 24: *Baldauf, M.*
Die effektive Rauhigkeit über komplexem Gelände – Ein Störungs-
theoretischer Ansatz.
Juni 1998

Nr. 25: *Noppel, H.*
Untersuchung des vertikalen Wärmetransports durch die Hangwind-
zirkulation auf regionaler Skala.
Dezember 1999

Nr. 26: *Kuntze, K.*
Vertikaler Austausch und chemische Umwandlung von Spurenstoffen
über topographisch gegliedertem Gelände.
Oktober 2001

Nr. 27: *Wilms-Grabe, W.*
Vierdimensionale Datenassimilation als Methode zur Kopplung zweier
verschiedenskaliger meteorologischer Modellsysteme.
Oktober 2001

Nr. 28:   *Grabe, F.*
Simulation der Wechselwirkung zwischen Atmosphäre, Vegetation und Erdoberfläche bei Verwendung unterschiedlicher Parametrisierungsansätze.
Januar 2002

Nr. 29:   *Riemer, N.*
Numerische Simulationen zur Wirkung des Aerosols auf die troposphärische Chemie und die Sichtweite.
Mai 2002

Nr. 30:   *Braun, F. J.*
Mesoskalige Modellierung der Bodenhydrologie.
Dezember 2002

Nr. 31:   *Kunz, M.*
Simulation von Starkniederschlägen mit langer Andauer über Mittelgebirgen.
März 2003

Nr. 32:   *Bäumer, D.*
Transport und chemische Umwandlung von Luftschadstoffen im Nahbereich von Autobahnen – numerische Simulationen.
Juni 2003

Nr. 33:   *Barthlott, C.*
Kohärente Wirbelstrukturen in der atmosphärischen Grenzschicht.
Juni 2003

Nr. 34:   *Wieser, A.*
Messung turbulenter Spurengasflüsse vom Flugzeug aus.
Januar 2005

Nr. 35:   *Blahak, U.*
Analyse des Extinktionseffektes bei Niederschlagsmessungen mit einem C-Band Radar anhand von Simulation und Messung.
Februar 2005

Nr. 36:   *Bertram, I.*
Bestimmung der Wasser- und Eismasse hochreichender konvektiver Wolken anhand von Radardaten, Modellergebnissen und konzeptioneller Betrachtungen.
Mai 2005

Nr. 37:   *Schmoeckel, J.*
Orographischer Einfluss auf die Strömung abgeleitet aus Sturmschäden im Schwarzwald während des Orkans „Lothar".
Mai 2006

Nr. 38: *Schmitt, C.*
Interannual Variability in Antarctic Sea Ice Motion: Interannuelle
Variabilität antarktischer Meereis-Drift.
Mai 2006

Nr. 39: *Hasel, M.*
Strukturmerkmale und Modelldarstellung der Konvektion über
Mittelgebirgen.
Juli 2006

Ab Band 40 erscheinen die Wissenschaftlichen Berichte des Instituts für Meteo-
rologie und Klimaforschung bei KIT Scientific Publishing (ISSN 0179-5619).
Die Bände sind unter www.ksp.kit.edu als PDF frei verfügbar oder als
Druckausgabe bestellbar.

Nr. 40: *Lux, R.*
Modellsimulationen zur Strömungsverstärkung von orographischen
Grundstrukturen bei Sturmsituationen. (2007)
ISBN 978-3-86644-140-8

Nr. 41: *Straub, W.*
Der Einfluss von Gebirgswellen auf die Initiierung und Entwicklung
konvektiver Wolken. (2008)
ISBN 978-3-86644-226-9

Nr. 42: *Meißner, C.*
High-resolution sensitivity studies with the regional climate model
COSMO-CLM. (2008)
ISBN 978-3-86644-228-3

Nr. 43: *Höpfner, M.*
Charakterisierung polarer stratosphärischer Wolken mittels hochauflö-
sender Infrarotspektroskopie. (2008)
ISBN 978-3-86644-294-8

Nr. 44: *Rings, J.*
Monitoring the water content evolution of dikes. (2009)
ISBN 978-3-86644-321-1

Nr. 45: *Riemer, M.*
Außertropische Umwandlung tropischer Wirbelstürme: Einfluss auf
das Strömungsmuster in den mittleren Breiten. (2012)
ISBN 978-3-86644-766-0

Nr. 46: *Anwender, D.*
Extratropical Transition in the Ensemble Prediction System of the ECMWF:
Case Studies and Experiments. (2012)
ISBN 978-3-86644-767-7

Nr. 47: *Rinke, R.*
Parametrisierung des Auswaschens von Aerosolpartikeln durch
Niederschlag. (2012)
ISBN 978-3-86644-768-4

Nr. 48: *Stanelle, T.*
Wechselwirkungen von Mineralstaubpartikeln mit thermodynamischen
und dynamischen Prozessen in der Atmosphäre über Westafrika. (2012)
ISBN 978-3-86644-769-1

Nr. 49: *Peters, T.*
Ableitung einer Beziehung zwischen der Radarreflektivität, der
Niederschlagsrate und weiteren aus Radardaten abgeleiteten
Parametern unter Verwendung von Methoden der multivariaten
Statistik. (2012)
ISBN 978-3-86644-323-5

Nr. 50: *Khodayar Pardo, S.*
High-resolution analysis of the initiation of deep convection forced
by boundary-layer processes. (2012)
ISBN 978-3-86644-770-7

Nr. 51: *Träumner, K.*
Einmischprozesse am Oberrand der konvektiven atmosphärischen
Grenzschicht. (2012)
ISBN 978-3-86644-771-4

Nr. 52: *Schwendike, J.*
Convection in an African Easterly Wave over West Africa and the
Eastern Atlantic: A Model Case Study of Hurricane Helene (2006)
and its Interaction with the Saharan Air Layer. (2012)
ISBN 978-3-86644-772-1

Nr. 53: *Lundgren, K.*
Direct Radiative Effects of Sea Salt on the Regional Scale. (2012)
ISBN 978-3-86644-773-8

Nr. 54: *Sasse, R.*
Analyse des regionalen atmosphärischen Wasserhaushalts unter
Verwendung von COSMO-Simulationen und GPS-Beobachtungen. (2012)
ISBN 978-3-86644-774-5

Nr. 55: *Grenzhäuser, J.*
Entwicklung neuartiger Mess- und Auswertungsstrategien für ein
scannendes Wolkenradar und deren Anwendungsbereiche. (2012)
ISBN 978-3-86644-775-2

Nr. 56: *Grams, C.*
Quantification of the downstream impact of extratropical transition
for Typhoon Jangmi and other case studies. (2013)
ISBN 978-3-86644-776-9

Nr. 57: *Keller, J.*
Diagnosing the Downstream Impact of Extratropical Transition
Using Multimodel Operational Ensemble Prediction Systems. (2013)
ISBN 978-3-86644-984-8

Nr. 58: *Mohr, S.*
Änderung des Gewitter- und Hagelpotentials im Klimawandel. (2013)
ISBN 978-3-86644-994-7

Nr. 59: *Puskeiler, M.*
Radarbasierte Analyse der Hagelgefährdung in Deutschland. (2013)
ISBN 978-3-7315-0028-5

Nr. 60: *Zeng, Y.*
Efficient Radar Forward Operator for Operational Data
Assimilation within the COSMO-model. (2014)
ISBN 978-3-7315-0128-2

Nr. 61: *Bangert, M. J.*
Interaction of Aerosol, Clouds, and Radiation on the Regional
Scale. (2014)
ISBN 978-3-7315-0123-7

Nr. 62: *Jerger, D.*
Radar Forward Operator for Verification of Cloud
Resolving Simulations within the COSMO Model. (2014)
ISBN 978-3-7315-0172-5

Nr. 63: *Maurer, V.*
Vorhersagbarkeit konvektiver Niederschläge:
Hochauflösende Ensemblesimulationen für Westafrika. (2014)
ISBN 978-3-7315-0189-3

Nr. 64: *Stawiarski, C.*
Optimizing Dual-Doppler Lidar Measurements of Surface
Layer Coherent Structures with Large-Eddy Simulations. (2014)
ISBN 978-3-7315-0197-8